Interstellar dust grain: diameter 4×10^{-6} inch

Blue light wavelength: 1.9×10^{-5} inch

Bacterium: diameter 4×10^{-5} inch

Black hole: diameter 40 miles

Large moon crater: diameter 120 miles

Largest asteroid: diameter 620 miles

Mars: diameter 4,217 miles

White dwarf: diameter 5,000 miles

Venus: diameter 7,521 miles

BETWEEN THE STARS

Aimed at the heart of star cluster NGC 2264, a dense cloud of interstellar dust called the Cone nebula cuts off light from stars behind it. The reddish areas are emission nebulae—atoms of hydrogen gas excited by ultraviolet radiation from the cluster's young, hot stars.

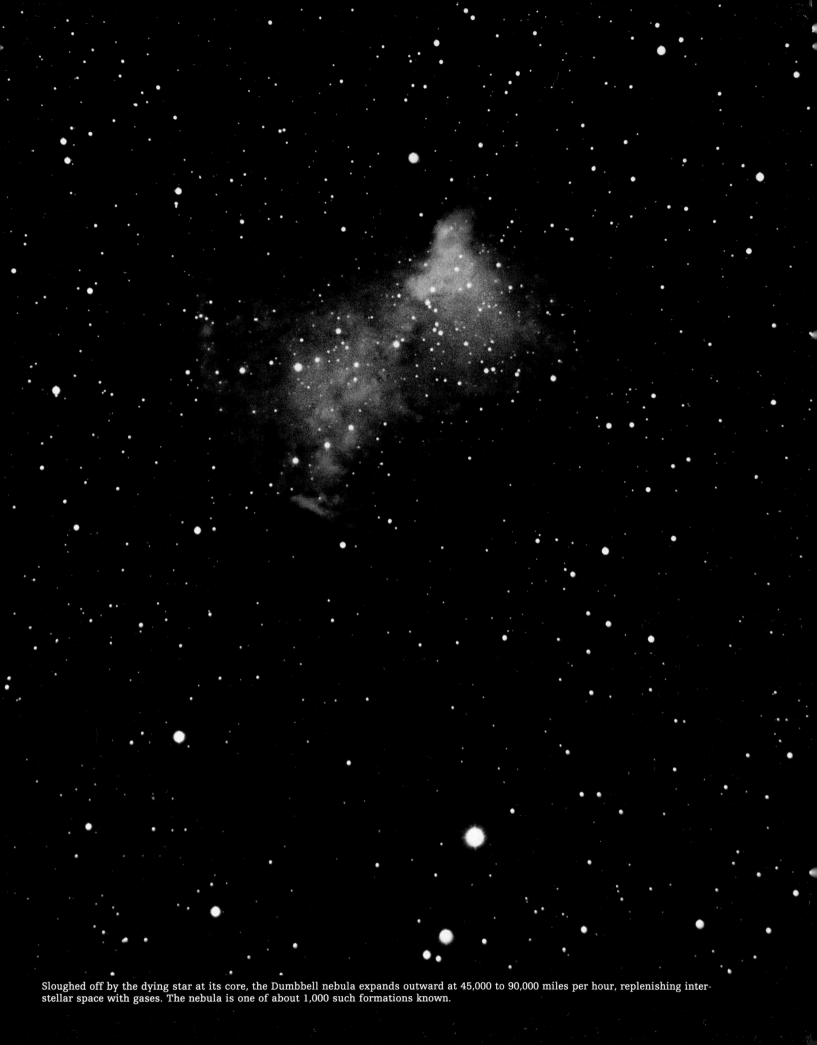

Sloughed off by the dying star at its core, the Dumbbell nebula expands outward at 45,000 to 90,000 miles per hour, replenishing inter-stellar space with gases. The nebula is one of about 1,000 such formations known.

Luminous filaments of hydrogen, helium, carbon, and oxygen forming the Veil nebula glow from collisions with interstellar dust and gas. Located 2,500 light-years from Earth, the nebula is the remnant of a star that went supernova more than 30,000 years ago.

Locked in a gravitational dance some 700 million light-years from Earth, the roughly 300 galaxies of the Hercules cluster undergo a constant exchange of matter with the intergalactic medium.

TIME® LIFE BOOKS

This volume is one of a series that examines the universe in all its aspects, from its beginnings in the Big Bang to the promise of space exploration.

VOYAGE THROUGH THE UNIVERSE

BETWEEN THE STARS

BY THE EDITORS OF TIME-LIFE BOOKS
ALEXANDRIA, VIRGINIA

CONTENTS

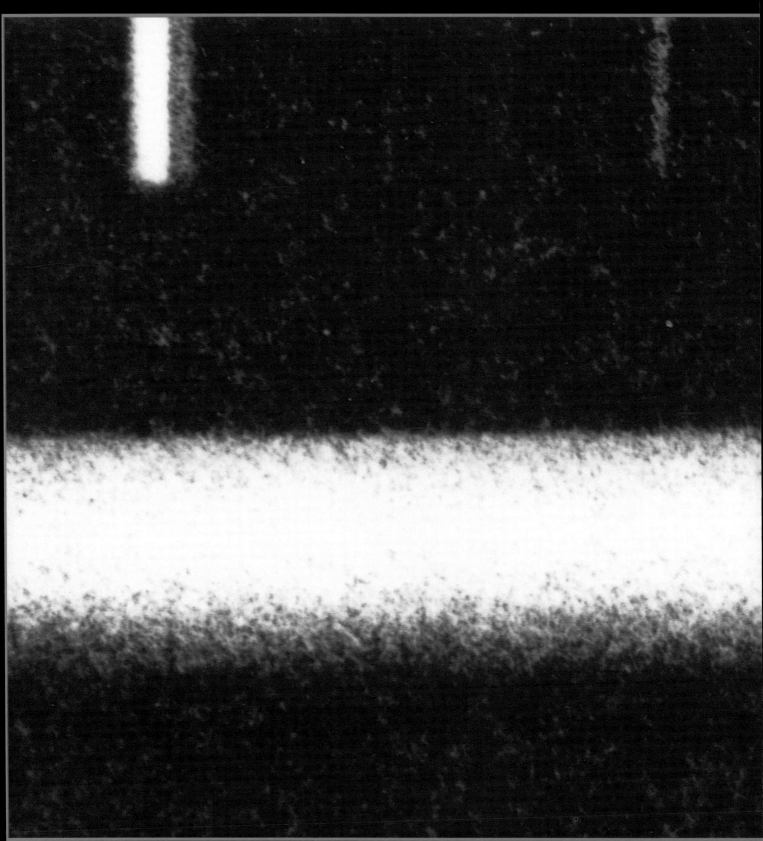

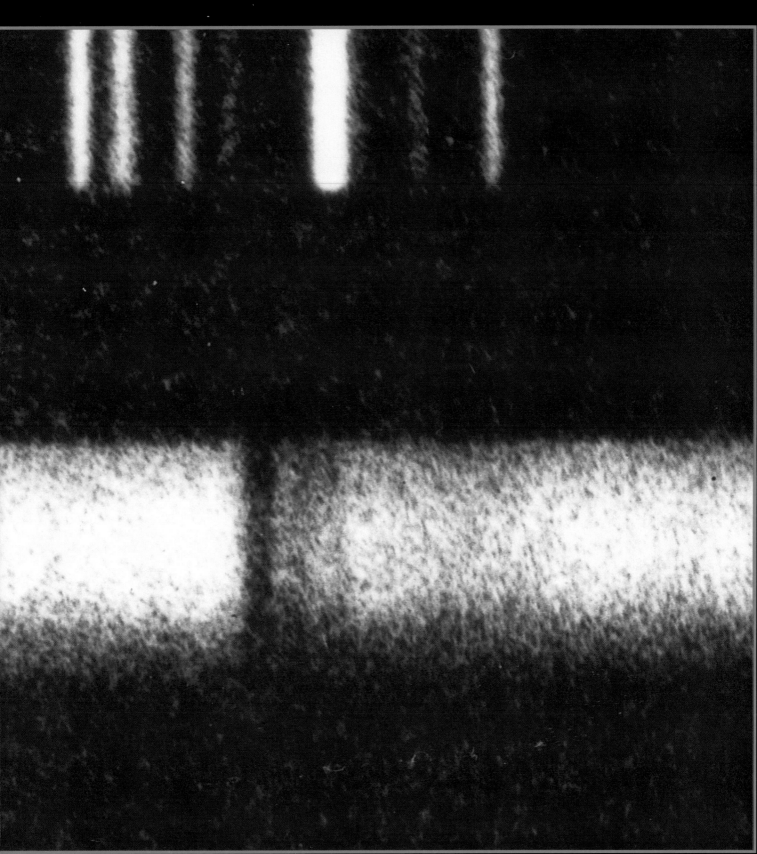

A telling mark in the stellar spectrum at bottom shows that a cloud of calcium—identified by the reference spectrum at top—has blocked some of the star's light. Such absorption lines helped confirm the existence of interstellar matter.

illiam Herschel certainly had no intention of stifling scientific inquiry in the latter part of the 1780s, when he formulated his "construction of the heavens" theories. In laying out his cosmological scheme, the great theorist, telescope maker, and discoverer of the planet Uranus imagined the evolution of an infinite universe, beginning with a uniform distribution of stars. Slowly over time, under the sway of gravity, this smooth fabric would be transformed into the universe as Herschel knew it—with stars gathered into distant congregations, such as the whitish patches known to observers as spiral nebulae. As he swung his telescope across the sky, the astronomer encountered numerous stretches that appeared totally blank, vast fields without even a single star in evidence. Herschel found these voids easy to account for: It was only natural, he wrote, that "there will be formed great cavities or vacancies by the retreat of the stars toward the various centers which attract them."

The renowned astronomer's view was splendid in its simplicity: Where there are stars, there is light; therefore, where there is no light, there is nothing. Herschel thus consigned enormous tracts of the universe to a kind of scientific limbo. Devoid of matter, they were also devoid of interest.

Like the pronouncements of so many great men, Herschel's hypothesis virtually had the force of law. For the next 150 years, scarcely anyone thought—or summoned the temerity—to investigate the "vacancies" that punctuated the universe. But the theory ultimately turned out to be wrong. And perhaps because of this false start, the quest to discover the true nature of those starless regions ranks as one of the most intriguing adventures in the annals of astronomy.

TUNING IN TO THE CELESTIAL SPHERE
Herschel's beliefs were based on what he could see through the optical telescopes of his day. But the visible light from stars represents only a relatively small fraction of the energy radiated from the sky above. The heavens broadcast across the entire electromagnetic spectrum: High-energy objects such as supernovae and pulsars emit short-wavelength energy in the form of gamma, x-, and ultraviolet rays. Stars and galaxies radiate much of their energy at the

midrange wavelengths of visible light, and cooler, low-energy objects such as planets and gaseous nebulae send out longer-wavelength infrared and radio waves. In the few decades since World War II, scientists have devised supersensitive radio telescopes and infrared, ultraviolet, x-ray, and gamma ray detectors to complement the optical instruments that had until then borne the burden of discovery. With these revolutionary new tools, astronomers have found, among many other things, that the space between the stars (William Herschel's age-old "vacancies") is actually filled with matter. In the Milky Way alone, that matter—enormous clouds consisting predominantly of gas and other vast clouds containing mainly dust—adds up to a mass equaling perhaps five percent of the mass of the stars themselves. Yet, in the immensity of space, this interstellar matter remains so sparse that, by comparison, the most perfect laboratory vacuum on Earth is as jam-packed and pulsing with energy as a hive full of bees *(page 16)*.

THE BOY WHO LOVED TELESCOPES

The first astronomer to provide substantial evidence for the existence of interstellar matter was Edward Emerson Barnard, a scientist so respectful of Herschel that he refused to his dying day to declare the earlier astronomer unequivocally wrong. Though born thirty-five years after Herschel's death in 1822, Barnard had much in common with his predecessor. Both began as amateurs. Both took inspiration from popular books; Herschel read about the discoveries of Newton, and Barnard in his turn read about the discoveries of Herschel. And both men possessed a lonely genius for observation and interpretation. A contemporary of Barnard's called him the keenest eye that ever looked through a telescope—with the possible exception of Herschel.

Barnard, whose father died before he was born, grew up in poverty in Nashville, Tennessee, raised by his invalid mother. The lad received a scant two months of formal education and was working at a job in a photo shop—Van Stavoren's Photographic Gallery—by the time he was nine. To provide the intense light necessary to make prints with the relatively insensitive photographic papers of the time, the shop used a huge enlargement camera, not unlike a telescope, aimed at the Sun. Lacking a clock drive to keep the device on the solar track as sunlight was beamed through the negative, Van

English astronomer William Herschel—shown here at a celestial sphere pointing to his most famous discovery, Uranus—declared in the 1780s that the dark regions of the Milky Way were "holes in the heavens," empty of matter. The astronomical community stood by his assertion for more than 100 years, until advances beyond the optical telescope proved him wrong.

MATTER IN THE VACUUM OF SPACE

In laboratories on Earth, scientists can create vacuums that are billions of times more tenuous than the atmosphere, but this achievement pales in comparison to the vacuum of interstellar space. The regions between the stars, inhabited by a smattering of gas atoms and molecules one ten-millionth the size of a pinhead and dust grains only a few hundred times larger, are typically a trillion times more desolate than even the best laboratory vacuums. A cubic centimeter of interstellar space—a volume the size of a small marble—holds scarcely one gas atom or molecule. Interstellar dust is an even rarer commodity: On average, only one particle would be found in a volume the size of a football stadium.

Yet over the unimaginable reaches of space, these far-flung bits of matter are capable of accumulating into formidable structures that obscure astronomers' views of the stars beyond them. A tube only one centimeter square stretching from the Sun across four light-years to Alpha Centauri would accommodate a million dust particles and a billion billion atoms of gas. Over even greater distances, enough gas and dust build up along the lines of sight from Earth to create the dark patches that astronomers once believed were simply holes in the heavens. Added across the entire 100,000-light-year diameter of the Milky Way, interstellar matter accounts for at least 10 percent of the galaxy's mass. Out of this vast ocean of galactic flotsam new stars are born, and into it dying stars cast their spent remains.

Stavoren employed neighborhood boys to turn geared wheels. The diligent young Barnard excelled and eventually won promotion at the establishment.

Four years of guiding the enlarger sparked an interest in lenses, along with a wish for a telescope that would give him a closer look at the stars and planets. The answer to Barnard's prayers came in the form of one J. W. Braid, an instrument maker and new employee at the gallery. Braid helped the boy find the parts for his first small instrument. "He was delighted," Braid later wrote. "The simple telescope gave Barnard more pleasure than anything else in his whole life." This joy, combined with his youthful experience in photography, would change astronomy forever. But for the moment, Barnard knew no astronomers, had no access to astronomy books, and had not even learned the names of the stars. Then a bit of luck came his way.

One day when Barnard was eighteen, an acquaintance showed up on his doorstep asking for money. Barnard had helped the man out before but had never been repaid. This time, however, the supplicant brought a book as collateral: *The Sidereal Heavens,* by Thomas Dick, a Scottish cleric and sometime astronomer. Within an hour, Barnard later wrote, he had learned the names of his old friends in the sky. Dick borrowed liberally from Herschel, and Barnard was entranced by descriptions of "self-luminous regions," now called bright nebulae, and the dark regions—the "holes in the heavens"—that Herschel supposed had been swept clean of matter during cluster formation.

At that moment, Edward Barnard embarked on what would become a life-long effort to make sense of the heavens. The pursuit led to a succession of bigger telescopes. In 1876, the nineteen-year-old bought an instrument with a five-inch aperture; it cost nearly $400, two-thirds of his annual wage, but served him well for years. The next year, he attended a meeting of the American Association for the Advancement of Science in Nashville, and was inspired to take up the search for comets, which observers were then studying with a vengeance. On May 12, 1881, he scored his first success, only to have it turn to ashes. Observing the area around the star Alpha Pegasi, Barnard came upon a very faint comet. He found it again the next night, but it was gone the night after, and no one else had seen it. Under international rules, more than one observer has to certify a discovery, and in any case, Barnard did not know how to make accurate plots of a comet's position in the sky. His comet sighting was rejected. Undaunted, the young man refined his skills and over the next six years located seven more comets. Properly certified and plotted, these finds were accepted. His growing abilities won him a position as assistant astronomer at Vanderbilt University in 1883, where he had access to a six-inch telescope. There he remained for four years, gaining practice and taking a variety of science courses. In 1887, he was invited to join the new Lick Observatory on Mount Hamilton in northern California.

This post proved more than a little frustrating. The observatory had a new and highly advanced thirty-six-inch refracting telescope, but Barnard was not allowed to use it for four years. Lick's autocratic director, Edward S. Holden, reserved the big instrument for himself and his senior assistants.

While biding his time and working with a much smaller instrument, Barnard turned his brilliance to pioneering the use of photography as a serious research tool in astronomy. The great advantage of photographic emulsions is their ability to accumulate light over several hours, enabling a patient astronomer to capture faint stars and nebulae impossible to see with the eye alone. The photographs Barnard made during this period inspired him to poetic description. The Pleiades, for example, were "filled with an entangling system of nebulous matter which seems to bind together the different stars with misty wreaths and streams of filmy light all of which is beyond the keenest vision and the most powerful telescopes."

Within these misty wreaths, Barnard also saw dark areas. A photograph of the area surrounding the star Theta Ophiuchi seemed to suggest, he wrote, that "these dark and black patches in it are thin places and actual holes." Herschel had written as much, and the common-sense idea still appealed to Barnard—as it did to most astronomers of the time.

One doubter was Arthur Cowper Ranyard, secretary of the British Royal Astronomical Society and editor of the popular magazine *Knowledge,* where Barnard described the Theta Ophiuchi photograph in 1894. A respected observer in his own right, Ranyard asserted in an editorial accompanying Barnard's photograph that for a dark patch to be a hole, it would have to be a tunnel through the bright nebula, pointing directly away from Earth. "The probabilities against such a radial arrangement with respect to the earth's place in space," Ranyard wrote, "[seem] to my mind to conclusively prove that the narrow dark spaces are due to streams of absorbing matter, rather than to holes or thin regions in bright nebulosity." To Ranyard, the dusky patches were masses of something that blocked the light of stellar regions behind them. He was right, of course, but scarcely anyone noticed at the time, and Barnard himself only acknowledged the insight many years afterward.

In 1892, Barnard finally gained access to the thirty-six-inch refractor and within two months he discovered the fifth moon of Jupiter, which skimmed so close to the planet's surface that it was nearly blotted out by the reflective glare. Later named Amalthea, after the goat that suckled the infant Zeus, the new moon was the first Jovian satellite to be found since Galileo's initial discovery of four moons in 1609, and it rocketed both Lick Observatory and Barnard to the pinnacle of astronomical success. Barnard worked at Lick for two more years, but he and director Holden bickered constantly, and Barnard never forgave the director for monopolizing the large telescope early on. When he received an invitation in 1894 to transfer to the University of Chicago's new Yerkes Observatory in Williams Bay, Wisconsin, Barnard hesitated hardly at all. The facility would be the site of much of his best work.

"A DULLNESS OF THE FIELD"

Yerkes was building a forty-inch refracting telescope, which would be the largest of its kind in the world. Meanwhile, Barnard was following up the trail of the mysterious dark patches, using whatever instrument was available. As he continued his observations of the region surrounding Theta Ophiuchi, he began flirting with the concept of obscuring matter in the skies. "Though there seemed to be scarcely any stars here, there yet appeared a dullness of the field as if the sky were covered with a thin veiling of dust," he wrote in 1897. Two years later, he suggested that the dark areas looked "as if the cavity were partly veiled with some sort of medium that itself had apertures in it." And in the constellation Scorpius, he suspected that "outlying whirls of this nebulosity have become dark, and that they are the cause of the obliteration of the small stars near."

Barnard was not alone in his investigations. Another detective was German astronomer Maximilian Wolf, a noted asteroid hunter who, in 1891, had devised a technique for discovering these bodies by the streaks they left on long-exposure photographic plates. Wolf went on to find 228 of them. But in 1904, he gave some consideration to whether a particular dark section in the constellation Cygnus was actually "a dark mass following the path of the nebula, absorbing the light of the fainter stars." Over the next several years, at Baden Observatory in Heidelberg, Wolf laboriously photographed and counted the stars in each class of magnitude, or brightness, in large segments of the sky. When he compared the number of stars for each magnitude in obscured regions with the number in adjacent unobscured regions, he found that the fainter stars all vanished in the dark regions, as if a curtain had dropped between the observer and the stars. Enormously intrigued, he would continue his painstaking investigations for nearly twenty years.

While Barnard continued to photograph bright and dark nebulae at Yerkes and Wolf counted stars in Heidelberg, still other astronomers looked for clues using a different technique—spectroscopy. Half a century earlier, scientists had discovered that the light of the Sun and stars, when passed through a narrow slit and then through a prism, produced a pattern of dark lines in an

Tennessee-born Edward Emerson Barnard had just discovered Jupiter's fifth moon when he shot this self-portrait at the eyepiece of the Lick Observatory's thirty-six-inch refractor in 1892. Barnard was the first astrophotographer to use long-exposure methods to film the dark regions of the Milky Way, capturing images of unprecedented clarity and detail *(pages 20-21)*.

otherwise continuous spectrum of colors; something in the stars was absorbing light at specific wavelengths. At about the same time, they learned that chemical elements heated to incandescence emitted a characteristic pattern of bright lines—known as an emission spectrum—and deduced that stellar absorption lines were simply the negative version of emission lines. Calcium, for example, emitted or absorbed wavelengths at 3,968 and 3,933 angstroms, the so-called H and K lines; sodium emitted wavelengths at two positions, 5,896 and 5,890 angstroms, known collectively as the D lines. (An angstrom is about four-billionths of an inch.) By matching the patterns of absorption lines in a stellar spectrum with the emission patterns of elements on Earth, scientists could determine what elements the stars were made of.

They further learned that the stellar absorption lines tended to shift toward one end of the spectrum or the other, depending on whether the star was moving toward or away from Earth, a phenomenon known as the Doppler effect. Wavelengths of light that are emitted from a source moving toward Earth would be blue-shifted, compressing to shorter wavelengths nearer the blue end of the visible spectrum; those emitted from a receding source would be red-shifted, stretching to longer wavelengths nearer the red end of the spectrum.

At about the time Wolf was beginning his star counts, German astronomer and instrument maker Johannes Hartmann of Potsdam Astrophysical Observatory discovered something curious about the absorption lines of the element calcium in the spectrum of a binary star called Delta Orionis in the constellation Orion. Hartmann knew the spectra of binary stars normally showed periodic Doppler shifts because of their complicated orbital motion. However, as he wrote in a seminal paper published in 1904, "the calcium line exhibits a very peculiar behavior." Unlike all the other lines in the stars' spectra, the H and K calcium lines were always, in Hartmann's words, "extraordinarily weak, but almost perfectly sharp"—meaning that the lines, though pale, were narrow and hard-edged rather than dark, broad, and fuzzy.

Furthermore, Hartmann added, "the calcium line does not share in the periodic displacements of the lines caused by the orbital motion of the star." Because the lines showed no Doppler shift, he reasoned that they were not produced by the binary system. And because they did not appear in every stellar spectrum, he also rejected the possibility that they were the result of absorption by Earth's atmosphere. The conclusion seemed obvious: "At some point in space in the line of sight between the Sun and Delta Orionis, there is a cloud which produces that absorption." Hartmann realized that he even possessed corroborating evidence: Three years earlier, he had discovered a similar phenomenon in the spectrum of the star Nova Persei, which was then undergoing what he described as "stormy processes." The spectrum displayed not only stationary calcium lines but stationary sodium lines as well.

Hartmann's observations met with silence, in part because of the sluggish scientific communication of the time. Barnard, for example, evidently was not aware of Hartmann's work, and neither was the highly respected Dutch theoretical astronomer Jacobus Cornelis Kapteyn, who in 1908 proposed the

Three plates from Barnard's *Photographic Atlas of Selected Regions of the Milky Way*, published in 1927, reveal the cloudlike nature of what had once been considered voids in the heavens. "We are not looking out into space through an opening in the Milky Way," Barnard wrote of the Theta Ophiuchi region *(above, left)*. He regarded the areas around the constellation Taurus *(center)* and Rho Ophiuchi *(right)* strong "proof of the existence of obscuring matter."

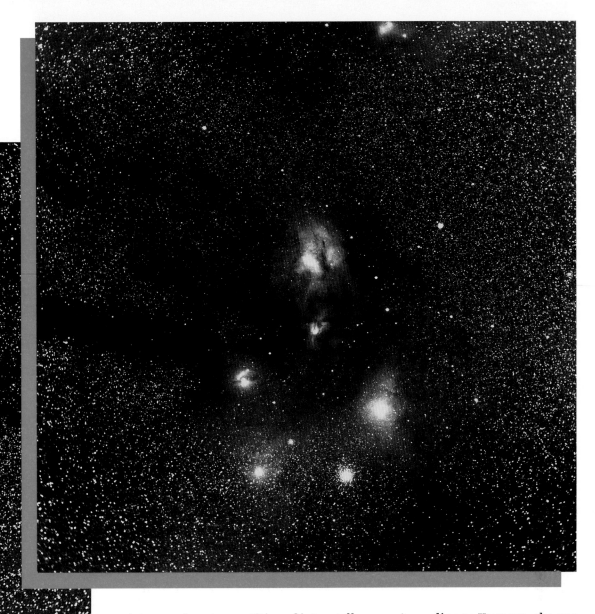

existence of vast quantities of interstellar gas. According to Kapteyn, the gas could be detected by what he called "space lines," absorption lines that "would not share in that part of the radial motion which is due to the motion of the stars themselves."

A GROWING INTEREST

Kapteyn's proposal was the first hint that scientific inertia might be giving way to movement. The next year, another spectroscopist took up the hunt. He was Edwin Frost, an American who, nearly twenty years earlier, had sharpened his skills during a semester's sojourn at Potsdam, where stellar spectroscopy was a major field of research. Frost's home base was Yerkes Observatory, which he headed from 1905 to 1932. Thus, he knew Barnard well but, inexplicably, does not appear to have compared notes with the older scientist on the subject of interstellar matter.

Perhaps because of his Potsdam connection, Frost did know of Hartmann's efforts, however. Like Hartmann, Frost combined the ability to design instruments with an uncanny scientific instinct. At Yerkes, he built a state-of-the-art spectrograph that was used with the observatory's forty-inch refractor, primarily to determine the radial velocities of stars—their speeds toward or away from Earth along the line of sight. With this device, Frost and others calculated the velocities of more than 400 stars, at the rate of about 10 a year. In April 1909, Frost made an important announcement. Two years earlier, he said, he had decided to reexamine all the spectrograms at Yerkes that contained calcium lines similar to those described by Hartmann. To his surprise, his review turned up twenty-five cases, about half of which were binaries. In all of them, the sharp calcium lines appeared relatively fixed in relation to the movements of the stars associated with them.

Further confirmation came at year's end from Vesto M. Slipher of the Lowell Observatory in Flagstaff, Arizona. Studying spectrograms from binary stars in the constellations Scorpius, Orion, and Perseus, Slipher too found sharp, stationary lines of calcium. "The same phenomenon," he wrote in the *Lowell Observatory Bulletin* in December, "exists in three widely separated regions of the sky." After analyzing the motion of the stars and that of the calcium lines, he concluded, as had Hartmann, that the lines must originate in interstellar space. However, he declared, "at present, it is not possible to decide whether the phenomenon is due to extensive veils covering large regions of the sky like the affected region of Scorpio, or whether the individual stars in the region each possess, perhaps more optically than physically, a nebulous veil whose chief absorption is, so far as known, calcium."

Slipher gingerly favored the assumption of "an interposing cloud covering at least certain extensive regions of the sky," even suggesting that calcium lines might be Kapteyn's space lines. But the concept seemed to stick in his throat. "It is also probably difficult to imagine the existence of a cloud of

THE EVIDENCE OF STARLIGHT

In the early 1900s, astronomers found revelations about the regions between stars in the runic lines of stellar spectra. Johannes Hartmann, studying a binary star system, stumbled on calcium absorption lines that remained fixed over time while the rest of the pattern shifted toward the red or blue end of the spectrum. Since such Doppler shifts result as the stars' orbits move them toward and away from Earth, he realized that he had detected an otherwise invisible calcium cloud somewhere along the line of sight to the system. Other scientists widened the spectral search to all types of stars, throughout the Milky Way. The stationary lines that appeared in their studies led to an irrefutable conclusion: Gas suffuses the galactic realms once thought to contain only nothingness.

1904 Sharp, fixed absorption lines in the spectrum of Delta Orionis led Johannes Hartmann to deduce that a cloud of calcium gas lay between Earth and the binary system.

1909 Edwin Frost, conducting a spectrographic review at the Yerkes Observatory, announced the discovery of stationary calcium lines in the spectra of twenty-five stars.

calcium vapor in stellar space," he wrote in the same paper, "because it seems incompatible with our present physical and chemical knowledge." Yet if Slipher was being cautious, his caution also went the other way: "We would hardly presume," he continued, "to think that knowledge is now complete."

In 1912, Slipher added another piece of evidence to the growing case for interstellar matter, but this one had nothing to do with stationary lines of gas. Studying a bright milky patch in the Pleiades, he noted that its spectrum exactly matched that of a nearby star. Since the chances of two objects having identical spectra were virtually impossible, Slipher reasoned that the nebula must be reflecting the star's light. And since the light was reflected exactly, with no new absorption lines to hint at the presence of other gases, he deduced that the nebula must consist of dust, or in Slipher's words, "pulverulent matter." At the time, however, his discovery fell on unreceptive ground. (Astronomers later classed bright clouds of this type as reflection nebulae.)

"BLACK AS DROPS OF INK"

The evidence was now largely at hand to explain the dark spaces in the sky. A year after Slipher's discovery of reflection nebulae, a chance observation led Edward Barnard to a significant insight. "I was struck by the presence of a group of tiny cumulus clouds scattered over the rich star-clouds of Sagittarius," he wrote later. "They were remarkable for their smallness and definite outlines—some being not larger than the moon." Noting that the earthly clouds were "as conspicuous and black as drops of ink" against the starry background, Barnard added that they were "in every way like the black spots shown in photographs of the Milky Way, some of which I was at the moment photographing." Barnard continued the train of thought: "One could not

1909 At the Lowell Observatory, Vesto Slipher found evidence of interstellar matter in three stellar spectra and suggested it was not associated with specific stars.

1923 John Stanley Plaskett, studying calcium and sodium lines in the spectra of forty stars of various types, across the galaxy, determined that interstellar gas is widespread.

1928 Otto Struve published measurements of stationary absorption lines for thousands of stars—evidence of an even distribution of gas throughout the Milky Way.

CRACKING THE SPECTRAL CODE

Scientists have gleaned almost everything they know about stars and the matter that lies between them from the painstaking examination of stellar spectra—patterns of dark and bright lines produced when light from an object passes through a finely divided grating.

An object's spectrum reveals not only its temperature, density, and velocity but also—and most important—its composition.

Every chemical element or molecule generates its own distinct spectral signature, and the key to deciphering it lies in the quantum-mechanical nature of the atom. The electrons in an atom surround the nucleus in a series of successively higher energy levels. In order for an electron to jump up to a higher level, it must absorb a photon whose energy is exactly equal

Emission. In the diagram at right of a hydrogen atom, an electron is boosted from the second to the sixth energy level, an event that can occur as a result of collisions with other atoms or the input of energy from an external source such as a star. When the electron drops back down to the second energy level *(center)*, it emits a photon of violet light. If this were the only transition taking place, the result would be a single emission line of violet light at 4,102 angstroms on the spectrum *(far right)*.

Continuous. In a hot, dense environment, such as deep within a star, every conceivable electron transition takes place in every atom and ion *(right)*. In theory, the result should be an emission spectrum of countless narrow, discrete lines, but in such surroundings, the energy levels within individual atoms are smeared, so that a given transition might not generate a photon of the same energy and color as it would in a cooler, less dense environment. Given the smeared levels and many transitions, very broad lines are produced, overlapping each other to yield a continuous spectrum *(far right)*.

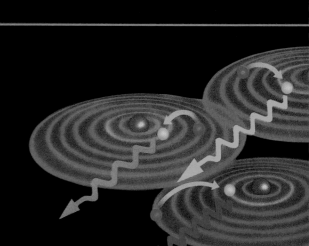

Absorption. As this continuous energy passes through the more diffuse stellar atmosphere, some of it is absorbed. At right, a hydrogen atom in the stellar atmosphere absorbs a photon of violet light which kicks its electron from the second to the sixth energy level. When many hydrogen atoms do this, the result is an absorption spectrum with a violet line removed *(far right)*. In a stellar atmosphere, absorptions from many different elements and molecules produce a spectrum banded by dark lines.

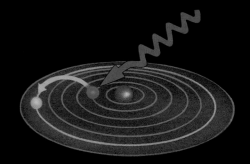

to the energy difference between the two levels. Conversely, for an electron to move down to a lower level, it must emit a photon equal in energy to the difference between the two levels. Since each element has a unique arrangement of energy levels, the precise energy—or particular wavelength of light—absorbed or emitted through a given electron transition is characteristic of that element.

The three basic types of spectra—emission, continuous, and absorption—are described below, using hydrogen as the source. The predominant component of most stars, hydrogen emits or absorbs photons of visible light at wavelengths between 3,900 and 7,500 angstroms when electrons drop to or jump from the second energy level. (An angstrom is about four-billionths of an inch.) In a laboratory, these transitions produce emission or absorption lines at specific positions on the spectrum. Shifts from these positions in a stellar spectrum reveal information about the motion of the star (page 26).

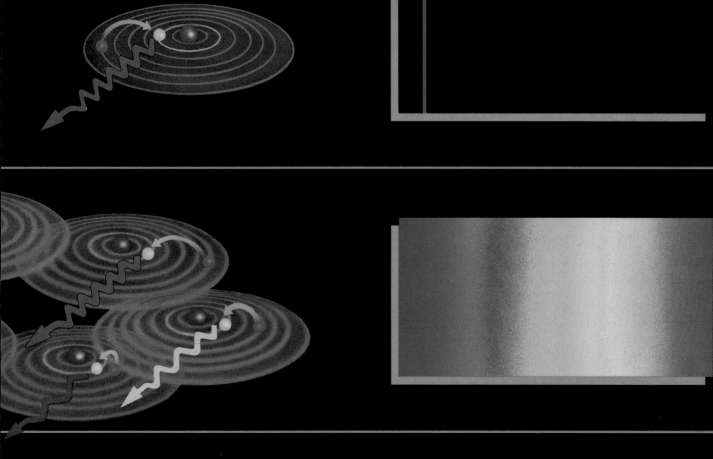

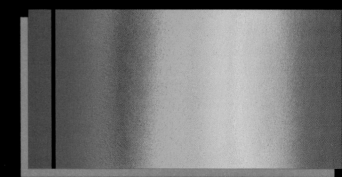

READING STATIONARY SIGNALS

Astronomers first detected interstellar gas when they observed that certain dark absorption lines in the spectra of binary star systems did not share features that were characteristic of the rest of the spectrum. A comparison of spectrograms made at two different times for the same binary system reveals the differences *(above)*.

In contrast to the broad, fuzzy lines of most of the spectrum, one line *(white arrow)* is narrow and sharp. In addition, the sharp line remains in the same place on the two spectrograms, while the others move as a result of the orbital motions of the two stars in the system, a phenomenon known as Doppler shifting. Spectroscopists thus reasoned that the stationary lines were generated not by the binary itself but by some intervening material.

resist the impression that many of the black spots in the Milky Way are due to a cause similar to that of the small black clouds mentioned above—that is, to more or less opaque masses between us and the Milky Way."

In 1913, he made his impressions known when he published "Photographs of the Milky Way and of Comets," a compendium that included one of the first photographs of the Horsehead nebula, a dark cloud in Orion that resembles the silhouette of the head and mane of a horse *(page 28)*. In describing the image, Barnard wrote, "One would not question for a moment that a real object—dusky-looking, but very feebly brighter than the sky—occupies the place of the spot. It would appear, therefore, that the object may not be a vacancy among the stars, but a more or less opaque body." Exactly what this opaque body consisted of, or how it came about, remained a mystery.

Six years later, in the January 1919 issue of the *Astrophysical Journal,* Barnard published a catalog of 182 dark nebulae; in the accompanying text, he struggled to reconcile his observations with Herschelian dogma. The paper was a monument to the difficulty of this struggle. He began with a caveat, still clinging to Herschel: "It would be unwise to assume that all the dark places shown on photographs of the sky are due to intervening opaque masses between us and the stars. In a considerable number of cases no other explanation seems possible, but some of them are doubtless only vacancies." Later, acknowledging for the first time the prescience of Arthur Ranyard, who twenty-five years previously had asserted that Barnard's "vacant spaces" looked to him like obscuring matter, Barnard cautiously cited his own work as evidence in favor of "obscuring bodies nearer to us than the distant stars," adding, "I think there is sufficient proof now to make this certain."

One semiconvert does not a revolution make, however. More widespread conversion would take time. In 1923—the year in which Edward Barnard "slipped quietly among the stars," as a friend so nicely phrased it—a Canadian spectroscopist picked up the calcium line trail that had cooled in the fourteen years since Edwin Frost and Vesto Slipher had worked with the spectra of binary and other stars.

John Stanley Plaskett, the director of the Dominion Astrophysical Observatory in Victoria, British Columbia, studied the spectra of forty stars of several types and in many places in the sky. He found that the Doppler shifts (and the radial speeds they represented) of the H and K calcium lines and the D lines of sodium differed dramatically from the shifts of the other lines in the stars' spectra. For example, six stars located near the constellation Perseus were moving toward Earth at roughly thirty miles per second, whereas whatever produced the calcium and sodium lines was approaching at only about twelve miles per second. As Frost had found, the velocity differences and the sharpness of the lines held true not just for binary stars but for all of the stars he studied. Moreover, when the Sun's own movement through space was subtracted, the calcium and sodium lines were effectively at rest.

The calcium and sodium lines, Plaskett later reported, must therefore be "produced in some uniform way independent of the widely different conditions prevalent over such a wide range of spectral types." Plaskett proposed that interstellar gas was widespread throughout space, and that as a star rushed through it, it ionized the calcium to produce the observed absorption lines. (As it turned out, Plaskett was essentially correct about the distribution of interstellar gas, but wrong as to how it was ionized.)

A FIRST OVERVIEW

Thus, in the course of three decades, Barnard, Wolf, Hartmann, Frost, Slipher, and Plaskett had assembled a wealth of evidence for the existence of interstellar matter, whatever its composition might ultimately prove to be. But it remained for a representative of a new generation of astronomers to turn the study of interstellar matter into one of the most exciting pursuits in astronomy. The scientist who now entered the field was England's Arthur Stanley Eddington, a towering presence in astrophysics who, in the mid-1920s, was at the peak of his career.

Soon after graduating from Cambridge University in 1905, Eddington had been named to the prestigious post of chief assistant at the Royal Observatory. Eight years later, at age thirty-one, he had returned to his alma mater to become Plumian professor of astronomy and director of the university observatory. In this capacity, he had led a 1919 expedition to West Africa to observe a solar eclipse and, not incidentally, to test a prediction made by Albert Einstein that light (such as that from a distant star) is bent by gravity as it passes a massive object (such as the Sun).

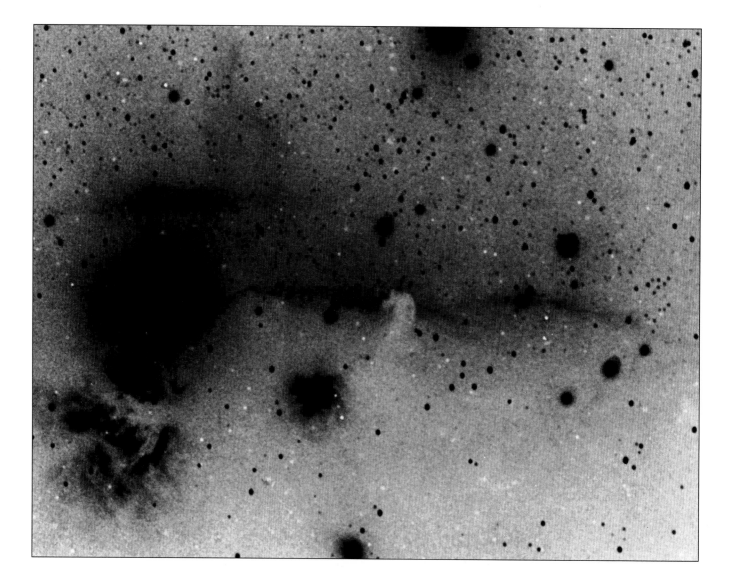

As one of the first to recognize the implications of Einstein's theory of relativity and the first to write an English-language explanation of it, Eddington effectively linked Britain to the revolutionary world of Continental physics. Astronomers paid close heed when in 1926 he chose "diffuse matter in space" as his topic for an annual lecture to the Royal Society in London. The idea of diffuse matter intrigued Eddington for several reasons. If light was being absorbed as it passed through space, many assumptions in astronomy would be up for grabs. Not least among the things needing revision would be distance estimates for a variety of celestial objects, all of which had been calculated without taking into account any dimming of light. Eddington also saw in interstellar matter a possible means of sustaining the processes of stellar evolution. He believed—incorrectly, as it later turned out—that stars rapidly lost mass as they burned. Therefore, they had to keep adding matter over their lifetimes or speedily die. Interstellar matter might provide the necessary supply of fuel, a supply that would be virtually inexhaustible.

Nestled among the belt stars of the constellation Orion, the Horsehead nebula rears its head and shoulders *(above, center)* in one of the first photographs ever to record the distinctive cloud. Edward Barnard, working at the Yerkes Observatory in 1913, detected the nebula and published this enlargement of his original plate as a negative in order to highlight details.

In his appearance before the Royal Society, Eddington presented the first overview of interstellar matter, in a way that would have shocked Edward Barnard. Without ado, he simply dismissed the Herschelian paradigm that Barnard had spent his career struggling to prop up. Astronomers must begin, Eddington said, by presuming "the existence of such matter unless or until its absence is proved." Citing as evidence the multiple findings of stationary calcium and sodium lines, he declared it probable that diffuse clouds of gas pervaded space. Having thus accepted the existence of interstellar gas, Eddington swiftly proceeded to calculate how much of it there was in the universe. Again based on the stationary line findings, he reckoned the density of interstellar gas at about ten hydrogen atoms per cubic centimeter and suggested that the gas represented "four to five times the mass aggregated into stars." To the audience, this was a startling notion. The prevailing view was that stars held the bulk of the universe's mass, and now Eddington had declared that the universe was instead made up of vast amounts of matter that was not only largely invisible but not even a reality to many astronomers.

Eddington then theorized that in a "normal region of space away from the preponderating influence of any one star," the temperature of interstellar matter was about 10,000 degrees Kelvin. Later evidence showed this estimate to be too high by several orders of magnitude except in the immediate vicinity of a star, but in explaining how the gas might attain such temperatures, Eddington suggested several legitimate mechanisms. For example, high-energy, short-wavelength ultraviolet radiation emitted by stars could heat the gas enough to knock electrons out of their orbits in atoms of neutral hydrogen (known as HI), thereby creating hydrogen ions (HII). Eddington later came to believe that neutral hydrogen in the interstellar medium would absorb most of the ionizing radiation emitted by stars before it could travel very far, so that most interstellar hydrogen should remain in its cooler, neutral form.

Next Eddington raised what he called a "grave difficulty"—how to account for "the obscuring power of dark nebulae." Given the extent to which such nebulae dimmed the light of stars behind them, he was compelled to wonder if the obscuring matter might take the form of "fine solid particles." He was not alone. Three years earlier, Max Wolf, as part of his long star-count studies, had published a paper entitled "On the Dark Nebula NGC 6960." Wolf concluded that the "light-capturing cloud" in the constellation Cygnus must consist "for the most part of a dust mass," giving two reasons for his belief. First, for interstellar gas to account for the observed dimming, it would have to be so dense that its total mass would affect stellar velocities, which is not the case. Second, in a phenomenon known as Rayleigh scattering, gas atoms would scatter light at the blue end of the spectrum, strongly reddening the light of faint stars shining through the gas as compared with the light of stars in unobscured regions; Wolf saw no such reddening. (As would be discovered later, dust particles also tend to weakly redden starlight, but in the mid-1920s astronomers did not yet know how to detect those effects.)

Troubling as Eddington found the idea of dust, he confessed he could come up with no other solution. When someone in the audience raised the possibility that light might be as effectively absorbed by interstellar molecules, Eddington found it equally problematic. Given the extremely low densities in space, he did not see how individual atoms could reliably and regularly encounter other atoms to form molecules. (Another decade or so would pass before interstellar molecules were discovered.) Left with the prospect of what he termed "meteoric matter," he finished on a rather plaintive note: "If we admit meteoric matter in order to explain the dark nebulae," he said, "there is no safeguard that other regions of the sky will be as transparent as is generally assumed."

COMMUNITY INSPIRATION

His uneasiness regarding dust aside, Eddington's lecture stood as a resounding endorsement of the view that vast gas clouds inhabited the regions between the stars, and it galvanized the astronomical community. Astronomers were then also responding to new studies on the differential rotation of stars in the Milky Way, which showed that stars nearer the galactic center travel around the center faster than do stars farther out. Equations developed in the galactic rotation studies allowed astronomers to determine the average distances of stars in a region simply from their radial velocities.

In 1928, a young Russian-American astronomer named Otto Struve used the formulas to add observational weight to Eddington's theorizing. Though born into a family of noted astronomers, Struve might never have followed in their footsteps had it not been for the generosity and concern of Yerkes Observatory's Edwin Frost. As a twenty-two-year-old lieutenant in the defeated Czarist Army in 1920, Struve barely escaped to Turkey, where he was living in poverty when a letter arrived from Frost, who had heard of his plight through Struve's widowed aunt. Frost offered the young man a position at Yerkes—and Struve's life's work was assured. Eight years later, Struve completed an extensive study of the stationary lines of calcium and sodium, comparing their intensity with the indicated recessional speed and distance from Earth of whatever was producing them. He found a direct relation: The intensity of the lines increased linearly as the distance increased—a clear indication that the absorbing gas was evenly distributed throughout the galaxy.

At about the same time, John Plaskett and colleague Joseph A. Pearce in British Columbia were winding up a com-

Renowned astrophysicist Arthur Stanley Eddington *(below)* galvanized London's Royal Society during a 1926 lecture when he summarily dismissed Herschel's concept of vacancies in space. Citing stationary calcium and sodium lines in stellar spectra, he claimed that clouds of gas pervaded the cosmos—and that they constituted the bulk of the universe's matter.

In 1930, the Lick Observatory's Robert J. Trumpler (above) concluded that interstellar matter was absorbing the light from so-called open star clusters, making them seem fainter and thus causing astronomers to miscalculate their distance from Earth. His work, which was based on studies of 100 clusters such as M67 (top), led to a downward revision of the estimated size of the Milky Way and to the acceptance of "interstellar extinction," the dimming of starlight by intervening dust.

prehensive study of the spectra of hot, bright, young stars, designated as O and B stars in the accepted classification scheme, with an eye to locating the source of the stationary lines. Using the equations from galactic rotation theory to determine distances to the stars and to the gas clouds responsible for the stationary lines, the pair found the clouds to be, on the average, halfway between Earth and the star. This conclusion, reported in 1930, effectively confirmed Eddington's prediction that gas pervaded the Milky Way and made a strong case against a rival hypothesis that interstellar gas was actually closely associated with stars rather than being more widely distributed.

Still up in the air, as it were, was the case for interstellar dust. Then, in 1930, a Lick Observatory astronomer inadvertently devised a test that would seal the verdict. Robert J. Trumpler had emigrated to the United States in 1915 from his native Switzerland after completing his education and a stint with the Swiss Geodetic Survey. The young man then joined the staff of the Allegheny Observatory in Pittsburgh, where he chose to study open star clusters, loose-knit groups of relatively young stars, which tend to congregate near the central plane of the Milky Way. And it was his struggle to measure the distances and diameters of such clusters more accurately that led to a clear demonstration of the existence of interstellar dust.

By this time, Max Wolf's findings had persuaded most astronomers of the existence of light-blocking discrete dust clouds in certain regions of the sky, and the stationary lines of calcium and sodium seemed irrefutable evidence for selective light absorption by diffuse clouds of interstellar gas. Many continued to assume that most of space was transparent, however, with no general absorption, or overall dimming, of starlight. But that assumption, as Eddington had pointed out, and as Robert Trumpler would discover for himself, affected all attempts to map the distribution of stars and to determine the size and shape of the Milky Way.

Trumpler began with a basic law of physics. The farther light travels from its source, the more it diffuses, or spreads. That diffusion reduces the amount of light falling on an observer anywhere along the line of travel. The inverse-square law stipulates that brightness, or apparent magnitude, is inversely proportional to the square of the distance from the source: A source three times as far away as an identical nearby source will seem nine times as dim—provided interstellar space is truly transparent. A relationship between a star's apparent magnitude and its so-called absolute magnitude—its hypothetical brightness at a standard distance of 32.6 light-years—allows astronomers to calculate its distance. (Since stars of a given spectral type have been found to have similar absolute magnitudes, astronomers can assign a star a mean absolute magnitude by examining its spectrum.)

From spectrographic analysis, Trumpler estimated the spectral types and corresponding absolute magnitudes of the brightest stars in nearby clusters. Then, by plugging the stars' absolute and apparent magnitudes into the appropriate formula, he obtained the distance to each cluster. To safeguard against errors, he next took the linear diameters of the clusters on photographic plates and the distances he had derived to compute the clusters' actual diameters. What he found was an apparent systematic error: The clusters were inexplicably growing larger the farther away they were.

After considering the possible sources of this anomaly, Trumpler concluded he had only two choices: "either to admit an actual change in the dimensions of open clusters with increasing distance, or to assume the existence of an absorption of light within our stellar system." With no logical or physical reason for a relationship between cluster size and distance, Trumpler decided that his findings were evidence of the interstellar absorption of light.

This dimming had a cascading effect on all calculations based on the inverse-square law: "If interstellar space is not perfectly transparent," Trumpler noted, "this law does not hold; the apparent brightness decreases more rapidly, our distance results are too large, and the error increases with the distance of the cluster. The linear diameters computed with these distance results are then also too large, and the error also progresses with distance."

Trumpler then went further: Speculating that the absorbing material "is related to interstellar calcium or to the diffuse nebulae which are also strongly concentrated to the galactic plane," he theorized that the absorption should vary by wavelength, thereby changing a star's color. By this time, astronomers knew that a star's color is dependent on its temperature, which in turn is associated with its spectral type, and they had been puzzled for some time by a discrepancy between the expected and actual colors of open star clusters. Now Trumpler showed that if he factored in an estimated percentage of absorption, the discrepancies disappeared.

His logic was inescapable and far-reaching. If, as he had so ably demonstrated, the central plane of the Milky Way was pervaded by a layer of light-absorbing matter, all previous estimates of the distances to various important celestial objects—to say nothing of the size of the galaxy itself and the Sun's location in it—were up for grabs. And once astronomers affirmed the existence of interstellar matter, they were suddenly free to wonder where it came from, what it was made of, and what role it plays in the hidden workings of the cosmos.

FROM A TINY GRAIN

Sky watchers had discovered in the 1890s that mysterious dark blobs obscure certain regions of space, but it was not until the 1930s that they succeeded in identifying the shrouding agents as clouds of interstellar dust. And only in the 1980s, with the integration of recent advances in observation, theory, and experimentation, were scientists able to sketch out the life cycle of an individual dust grain. That process, illustrated on the pages that follow, hinges on a symbiotic relationship between particles of dust and the very stars whose light they prevent from reaching Earth.

Gregarious by virtue of self-gravity, interstellar dust grains tend to clump together in huge, turbulent, chemically active clouds. These clouds are created and constantly replenished by matter thrown off from dying stars, be they gracefully expiring red giants or violently exploding supernovae. The dust clouds are cosmic recycling facilities in which new stars coalesce from the ashes of their ancestors.

The diameter of most dust grains is less than one micron—that is, one-millionth of a meter—and some may be just .002 micron across. Nevertheless, these minuscule bits of matter provide ideal host sites for chemical reactions. Gas atoms spewed from neighboring stars surround the grains and cling to their surfaces, where they combine and recombine to form dozens of simple and complex molecules—in particular, many of the carbon-based compounds that are crucial to the emergence of life itself. Thanks to this unceasing interplay of stars and dust, the galaxy's open spaces harbor nearly 100 different kinds of molecules that were nonexistent in the aftermath of the Big Bang.

THE PROGENY OF STARS

Dust grains are the chemical building blocks of most celestial matter, yet they were absent at creation. That instant, the Big Bang, produced only the lightest, simplest elements: about 75 percent hydrogen, 24 percent helium, and trace amounts of lithium. These gases eventually coalesced to form the first stars.

Many first-generation stars were 300 times the mass of the Sun and a million times more luminous, so they tended to burn out within just 10 million years or so. During that time, gravitational pressure in the stars' cores crushed together the nuclei of hydrogen atoms, fusing them to form helium and releasing great quantities of heat, which counterbalanced the inward pressure of gravity. As the stars consumed their stores of hydrogen, the heat abated, and the stars began to collapse upon themselves, creating higher and higher core pressures—and manufacturing new elements in the process. The helium nuclei fused into carbon, the carbon into oxygen, and so on through progressively heavier neon, magnesium, silicon, phosphorus, chlorine, calcium, titanium, and finally iron atoms.

During their last few moments of gravitational collapse, the massive early stars underwent fusion reactions so powerful that the stars were torn apart by supernova explosions *(right)*. Rich in heavy elements, the ejected stellar gases moved out through space, where they ultimately cooled and contracted to form infinitesimal dust-grain cores *(pages 36-37)*. These cores probably measured just five to twenty nanometers (billionths of a meter) in diameter, and they most likely consisted of graphite, heavy metals such as iron, and an abundance of silicates—extremely stable compounds made up of silicon and oxygen.

A first-generation star ends its short but productive life in a dramatic supernova explosion. Billowing outward from the detonation are gases laden with carbon, oxygen, and silicon atoms—elements that were synthesized by thermonuclear fusion inside the star. As the atoms cool, they begin to clump together *(inset);* once the atoms in a cluster reach 50 to 100 in number, they interact to such a degree that they behave like a solid—an individual grain of dust. The dust grain will in turn form the raw material of future stars and planets.

A Matter of Molecules

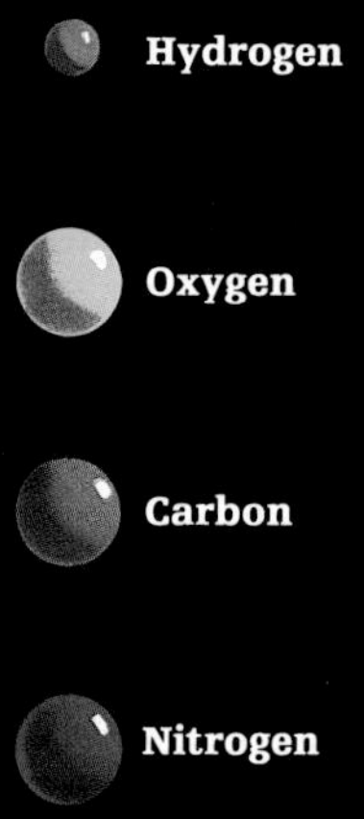

As a newborn speck of dust travels away from its parent star, it encounters a few simple elements that transform it into an object of remarkable complexity. Atoms of hydrogen, oxygen, carbon, and nitrogen—the most abundant components of the interstellar medium—adhere to the surface of the speck, where they combine to form various molecules. (Deprived of such a site for molecular mating, free-floating atoms are so widely dispersed through space that they rarely meet, much less bond; they constitute 99 percent of the interstellar medium.)

The compounds that initially form on the speck's surface are pictured on the opposite page; they are molecular hydrogen (made up of two atoms), water (three atoms), ammonia (four atoms), formaldehyde (four atoms), and methane (five atoms). As the molecules accumulate, they swathe the dust grain in a mantle of many layers. Every tier is a chemically dynamic site: Bombarded by the ultraviolet radiation that issues from stars and pervades interstellar space, the molecules constantly split apart and recombine to produce a variety of new compounds. Occasionally this bath of ultraviolet light causes a dust grain to lose matter, ejecting a spray of complex molecules that eventually attach themselves to other interstellar grains.

Unable to gather interstellar dust grains for laboratory examination, astronomers have inferred the objects' size and chemical composition by noting how clouds of interstellar dust scatter, absorb, and polarize the light from distant stars. Theories about the growth and evolution of dust grains continue to be tested in laboratories today, where scientists are staging chemical reactions in icy vacuum chambers that simulate the unimaginably frigid and desolate conditions of interstellar space.

Hydrogen

Oxygen

Carbon

Nitrogen

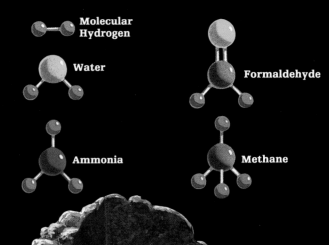

Molecular Hydrogen

Water

Formaldehyde

Ammonia

Methane

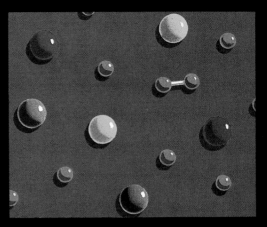

An interstellar dust grain begins to grow when atoms of hydrogen, oxygen, carbon, and nitrogen accumulate on the surface of a newly formed dust-grain core, colored gray in the large illustration (below, left). There, the atoms start to interact: Pairs of hydrogen atoms, for example, combine to form molecular hydrogen (left).

As additional atoms and molecules collect on the core, they form bonds that yield the simple chemical compounds shown at left: water (H_2O), ammonia (NH_3), and methane (CH_4). Because interstellar temperatures average only a few degrees above absolute zero, these compounds exist in a frozen state, sheathing the core in a mantle of mixed ices (blue).

Ultraviolet radiation (purple arrow) from nearby stars strikes the ice mantle, breaking its molecules into radicals—molecular fragments with a strong tendency to combine with one another. As the radicals interact, they produce formaldehyde (CH_2O) and other, more complex molecules that collect in a second mantle (yellow) of organic material surrounding the core. The combining radicals generate heat, which hurls random compounds such as methane into space (left).

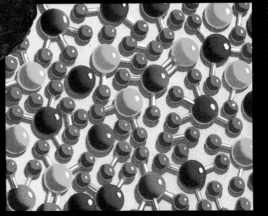

Dust grains that reside in particularly dense clouds or those that are subject to ultraviolet radiation for more than about a few million years may acquire yet a third mantle (red). Made up of frozen molecular hydrogen, water, ammonia, formaldehyde, and methane, this outer mantle can bring the dust grain's diameter up to almost one entire micron—nearly the size of a particle of smoke.

TRANSFORMED IN A STELLAR CRUCIBLE

Newly formed dust grains float through space for millions of years before being incorporated into vast, cold clouds of dust and gas, which then serve as sites for the creation of new stars. With temperatures below 10 degrees Kelvin and distances from end to end measuring as much as several hundred light-years, these clouds may contain as few as 50 particles per cubic centimeter or as many as 10,000. Their masses range from 0.1 to more than 500,000 times that of the Sun.

1

Buffeted by one of several possible forces—typically, the shock front from a supernova—an interstellar cloud of dust and gas *(right)* develops several high-density regions known as globules. The dust grains in each globule start to fall toward the globule's center; as a result, the core reaches a density threshold that causes the globule to undergo gravitational collapse.

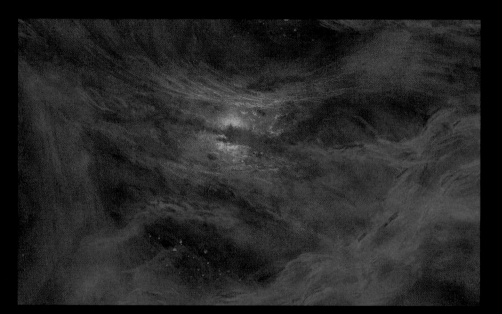

2

In this closeup view of a globule's collapse, in-falling particles of gas and dust *(yellow arrows)* add to the globule's mass, strengthening its gravitational field and enabling it to attract even more matter from the interstellar cloud. Dust grains distant enough to escape the pull of gravity form a dark cocoon nebula around the globule, screening the star-formation process from optical observation. To pierce this veil, astronomers use infrared detectors, which measure the heat thrown off by particles of dust and gas that collide during the collapse.

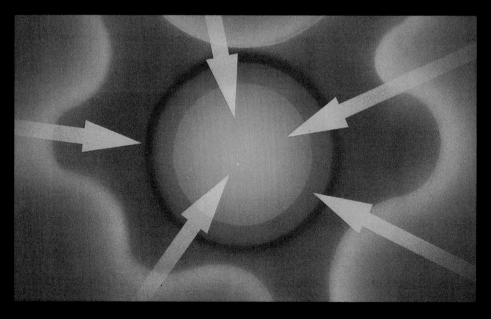

No matter what its density or mass, an interstellar cloud is subject to an array of random forces. It may collide with another cloud, or it may be struck by shock waves from a supernova explosion; the cloud may even be compressed by a density wave—a large-scale disturbance that moves through the gravitational field of the galaxy itself. Whatever the cause, the result is that portions of the cloud called globules attain locally high densities and begin to contract, eventually collapsing inward and forming protostars.

A protostar, for its part, may continue to collapse for a few hundred thousand to a few million years before its core reaches the internal pressure and temperature at which nuclear fusion begins. Roughly concurrent with that event, jets of material begin to shoot out of the north and south poles of the protostar, dispersing just enough of the enveloping cloud to expose the stellar progeny within.

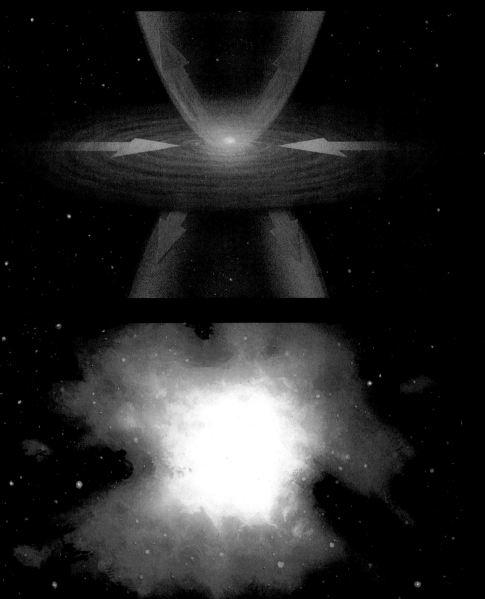

3

Typical of the protostars that are born inside a contracting globule is the one depicted at left. Like almost every protostar in this early stage of formation, it is girded by a thin, turbulent disk made up of dust and gas that are still accreting *(yellow arrows)* onto the object's surface. For reasons that continue to mystify astronomers, some of this accreting material is flung back into space; unable to make headway against the circumstellar disk, the outflow is channeled into powerful bipolar jets *(orange arrows)* that blow away portions of the surrounding cloud.

4

The contraction of the protostar raises its internal temperature to about 15 million degrees Kelvin, and thermonuclear fusion begins. The outward pressure of the fusion-generated heat counterbalances the inward pull of gravity; the protostar has now become a star, cloaked in a hot, tenuous nebula that takes either of two forms. If the imbedded star is hotter than about 30,000 degrees Kelvin, the star radiates so much ultraviolet light that the surrounding gas becomes ionized and glows, and the result is an emission nebula. If the star is cooler than about 20,000 degrees Kelvin, however, the gas will not be ionized, and the predominant source of light will be a reflection nebula *(left)*—a cooler mix of dust that glows by reflect-

RESTOCKING THE INTERSTELLAR LARDER

Just as the interstellar medium contributes dust grains to the creation of new stars, so too do stars constantly replenish their surroundings with all the elements necessary to form new dust grains. The supply line takes the form of a steady outflow of charged particles from every star in the galaxy; this stellar wind, as it is called, enriches the space between the stars with the by-products of nucleosynthesis.

The amount of dust-building material that a star contributes to the interstellar medium depends on both its mass and its mode of death. A large star—one of more than about four solar masses—perishes in a cataclysmic supernova explosion that hurls billions of tons of freshly synthesized elements into space. Among the ejecta is every element from helium through iron, as well as a number of heavy, neutron-rich elements—notably lead, plutonium, and uranium—created by the intense temperature, density, and radiation that attend the star's death.

A smaller star, comparable in size to the Sun, expires with a whimper rather than a bang, yet it contributes no less mightily to the interstellar medium. As the reserves of hydrogen fuel at its core run low, the small star begins to contract; this heats the object from within, causing its outer layers to expand until the star becomes a bloated red giant. As shown at right, the red giant then begins to shed concentric shells of stellar matter. Driven outward by increasing heat, the layers eventually merge to form a planetary nebula—so called because the rounded nebula appeared planetlike when viewed through the low-resolution optical telescopes of the mid-1800s. The core of the small star dwindles to form a white dwarf; the planetary nebula, meanwhile, adds its constituent matter to the interstellar medium, where some of it will clump together in dust grains that begin the cycle anew.

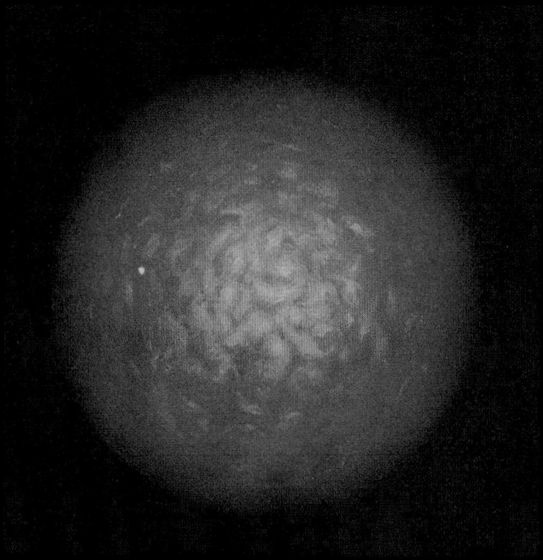

A dying star becomes a red giant *(above)* when its rising internal temperature causes it to slough off an outer layer of hydrogen gas. This huge shell moves outward from the star's surface, dispersing, decelerating, and cooling off as it travels. Mounting core pressures and temperatures then precipitate the high-speed ejection of a second layer of matter—a hot, rapidly expanding bubble that eventually catches up with the first shell and bursts through it. Stellar molting yields multiple ejected layers, each new one overtaking its predecessors to create a planetary nebula *(right)* laced with heavy elements that will cool to form a new generation of dust.

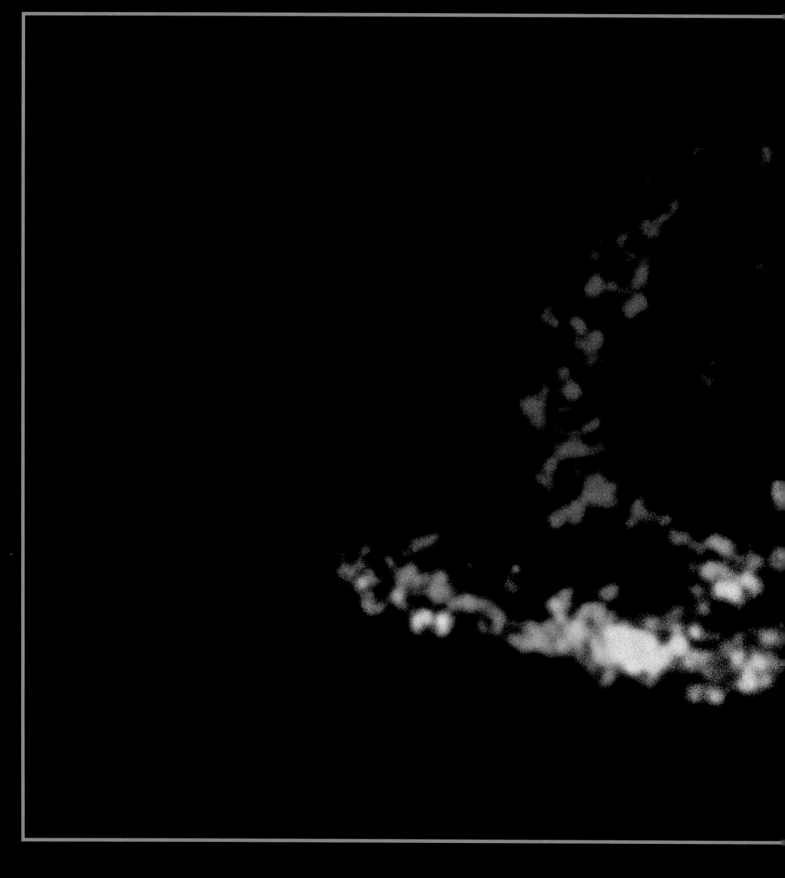

MAGNETISM

The sweeping spiral arms of the Whirlpool galaxy stand out in radio emissions given off by neutral atomic hydrogen (HI)—the main ingredient of interstellar space. The emissions here are color coded to reflect the Doppler shift caused by the galaxy's counterclockwise rotation: Approaching areas are blue, and receding areas are red.

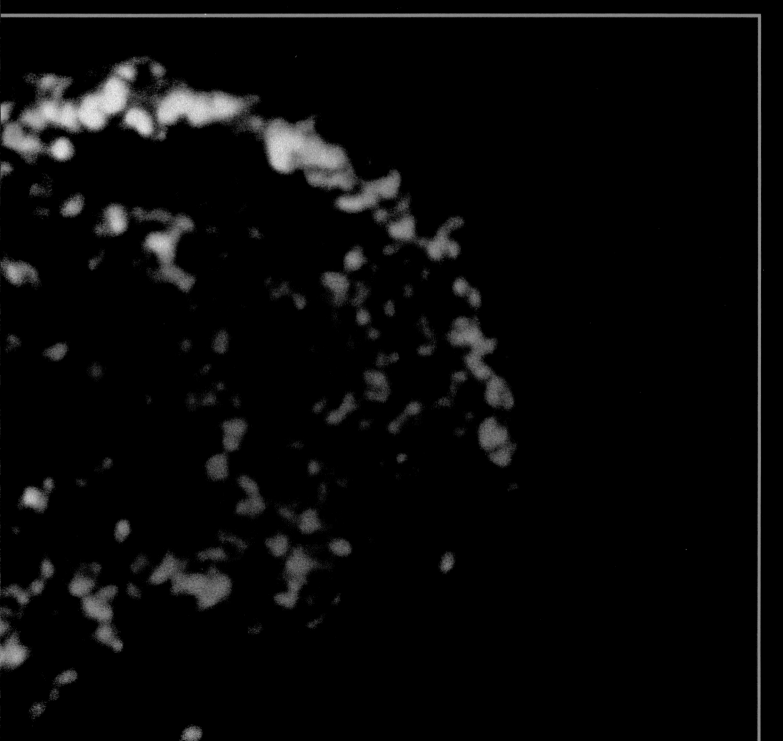

Robert Trumpler's 1930 report on the dimming of light from open star clusters near the midplane of the Milky Way was like the single loose pebble that started an avalanche. After years of practically ignoring all clues that interstellar space was not the transparent place it was long believed to be, astronomers seemed suddenly bent on making up for lost time. Once they had been persuaded that the apparent emptiness of space held solid grains as well as diffuse gas, they began trying to determine what the particles were made of, how they were formed, and precisely how they affected light.

The work built on—and, in turn, contributed to—scientific understanding of interstellar gas. Researchers realized that the stationary lines of calcium and sodium, though valuable in signaling the presence of those gases in the space between the stars, were inadequate for sketching a complete portrait of the galaxy's structure. Hydrogen, the most abundant element in the universe, was clearly the key to this endeavor, but detecting it in its most common, neutral form would require new tools. Over the next forty years, and especially after the advent of radio astronomy at midcentury, scientists would not only begin to plot the distribution of interstellar hydrogen but also discover a theretofore unsuspected phenomenon: the galactic magnetic field, whose effects on the interstellar medium and on starlight itself would help astronomers refine their portrait of the unassuming interstellar dust grain.

GROWING DUST GRAINS

When Swedish astronomer Bertil Lindblad, director of the Stockholm Observatory, took up the study of interstellar dust in 1935, he did so from a long interest in the composition and evolution of stars and stellar systems. Ten years earlier, Lindblad had achieved international prominence with detailed stellar velocity studies suggesting that stars in the Milky Way rotate around the galactic center: Some follow circular orbits flattened to the plane of the galaxy; others travel on more elliptical and inclined paths that loop high above and below the galactic plane. (Later refinements of these theories aided Otto Struve in his investigation of the stationary absorption lines produced by clouds of interstellar calcium and sodium.)

In turning to this new work, Lindblad began with several assumptions. First, he assumed that interstellar gas contains all the elements in approximately the same proportions as found in the Sun. Next, he accepted the reasoning, which had been put forth by Arthur Eddington in 1926, that emissions of high-energy ultraviolet radiation from stars ionized interstellar hydrogen, heating the gas—even at great distances from the emitting stars—to a temperature of 10,000 degrees Kelvin.

Finally, Lindblad assumed that solid particles would be warmed to a temperature of just three degrees above absolute zero by incident starlight in the galaxy. Examining the mechanics of energy transfer, he theorized that as hot gas atoms collided with cold particles, the energy of impact would be "rapidly radiated into space, or perhaps to some small extent transformed into subatomic energy, so that the particle remains cold." Thus, a particle could grow by accumulating additional mass, in the form of gas atoms, just as raindrops form in terrestrial clouds when water vapor condenses around dust and other airborne particles. Eventually—a billion years was Lindblad's estimate—the accretions would reach a size that matched the apparent size of the particles in the light-absorbing regions in the Milky Way. (Based on the absorption effects on visible light, which ranges in wavelength from .0007 to .0004 millimeter, scientists had recently estimated that the obscuring clouds contained particles on the order of a tenth of a millimeter in diameter.) Moreover, Lindblad noted, like interstellar gas, the dust would tend to follow the rotation pattern of stars around the center of the galaxy and would thus tend to concentrate in the galactic plane—a hypothesis that dovetailed precisely

1912 After a series of balloon flights in which he detected ionizing radiation increasing with altitude, Austrian physicist Victor Franz Hess concluded that such radiation must originate in space.

1925 American physicist Robert Millikan confirmed Hess's findings and dubbed the phenomenon "cosmic rays." Later research proved these rays were not radiation at all but streams of charged particles.

Unveilers of the Void

Throughout the twentieth century, astronomers have continued to refine their understanding of the interstellar medium, with each generation of investigators adding new details to the complex picture. Early efforts relied primarily on indirect evidence, such as the atmospheric effects of cosmic rays and the dimming of starlight by nebulae of dust and gas. Radio astronomy eventually made possible the direct measurement of interstellar atomic hydrogen and the subsequent identification of an interstellar magnetic field. All along, theorists have played a crucial role, their speculations on the medium's characteristics frequently guiding observers to new discoveries.

with Robert Trumpler's finding in 1930 of greater obscuration toward the midplane of the Milky Way.

Carrying his theory further, Lindblad also suggested that the gas in what he described as the "nebulous envelopes" observed around many young stars could evolve into particles even larger than those of smoke. In these high-density regions, the particles could, over the same time period, combine to form bodies the size of small asteroids. Once that happened, a kind of snow-ball effect would ensue, and growth would speed up. "It appears, therefore," he wrote, "that the net result of the condensation process will ultimately be a rapid growth of a comparatively small number of large bodies"—by which he meant planets and their moons.

Reasonable as the theory seemed, one hitch remained: Although astronomers were prepared to accept condensation as a likely mechanism for the growth of dust grains, no one could explain how the initial tiny cores, or nucleation sites, for the condensation process came into existence. Proof that it was possible would not be found for several decades.

OBSCURATION REVEALED

Even as Bertil Lindblad was delineating the mechanics of dust formation, American astronomer Alfred Joy, the son of a New England merchant and a former professor at the American University of Beirut, was winding up a long-term investigation of galactic rotation. In the process, he also recon-

1935 Swedish astronomer Bertil Lindblad postulated that grains of dust in the interstellar medium form and grow through the condensation of hot hydrogen gas around cold solid particles.

1939 Refining earlier studies, American astronomer Alfred Joy concluded that dust and gas are unevenly distributed in the Milky Way, with the highest concentrations in the vicinity of the galactic plane.

1939 Swedish astronomer Bengt Strömgren published a report demonstrating that while most interstellar hydrogen is neutral, distinct spheres of ionized gas surround hot stars.

1944 Dean of Dutch astronomers, Jan Oort suggested that the study of radio waves from space would yield critical information about the nature of interstellar matter and the structure of the galaxy.

firmed Trumpler's discovery that starlight is significantly dimmed in the direction of the galactic center.

Working with the 100-inch telescope at California's Mount Wilson Observatory, Joy had spent years plotting the motions, distances, magnitudes, and spectral types of thousands of stars. He was especially interested in a class of stars known as Cepheid variables, whose brightness fluctuates at a fixed, predictable rate. Because of an observed relationship between the radiance and fluctuation rates of Cepheids, astronomers could use these stars as convenient distance yardsticks. But the yardstick would be accurate only if the effects of interstellar light extinction—that is, absorption and scattering—were taken into account.

Joy observed 156 Cepheid variables, most of them located close to the equatorial plane of the galaxy. For purposes of comparison, he divided the stars into four groups, based on their distance from the Sun in all directions observable from the Northern Hemisphere, and plotted their radial velocities toward or away from Earth. Earlier work by Lindblad and others led Joy to expect these speeds to vary with the stars' distance from the galactic center. However, just as Robert Trumpler had found an anomalous relationship between distance and size in his study of open clusters, Joy discovered that stars did not travel at their predicted velocities.

He then considered the effects of light extinction, but even when he factored in an absorption rate of .85 magnitude per kiloparsec—meaning that a star's

1945 Setting the stage for later investigations, Hendrik van de Hulst, a student of Oort's, theorized that interstellar hydrogen ought to emit radiation at the radio wavelength of 21 centimeters.

1947 Dutch astrophysicist Bart Jan Bok proposed that small, round nebulae—Bok globules—were actually protostars, concentrations of dust and gas in the process of gravitational collapse.

1950 American astronomer Lyman Spitzer helped clarify the dynamics of the interstellar medium by describing mechanisms for heating and cooling both neutral and ionized hydrogen.

apparent brightness would decrease by 85 percent of a magnitude for each 3,260 light-years of distance from Earth—the discrepancy did not disappear. The stars in Group 4, the most distant group, had about the same radial velocities as those in the next most distant. "Either there is a breaking-down of the rotation effect at great distances in a way which cannot be accounted for," Joy concluded in a paper published in 1939, "or the absorption coefficient is greater for the stars of Group 4." And since there was no obvious reason why rotation effects should break down with distance from Earth, it was far more likely that the stars in Group 4 were located "in regions of greater obscuration than those of the nearer groups."

As it happened, recent studies of stationary lines of calcium and sodium seemed to indicate that these interstellar gases were not uniformly distributed in space; Joy suggested that the same might hold true for dust clouds. "In the existing state of our knowledge," he wrote, "it may not be possible to set a definitive value for the absorption throughout the galaxy. The results from one group of stars may well be quite different from [those] given by another group, depending on their positions and distances." In some places, Joy theorized, interstellar dust might be dense enough to produce an absorption coefficient as high as 1.5 magnitudes per kiloparsec, and perhaps even higher near the galaxy's midplane.

The question of the distribution—uniform or otherwise—of interstellar matter surfaced repeatedly during this period, particularly in connection

1951 Edward Purcell *(below, left)* and Harold Ewen, both of Harvard University, became the first astronomers to observe the 21-centimeter line for interstellar hydrogen on an emission spectrum.

1951 Americans Jesse Greenstein *(below, left)* and Leverett Davis proposed that the polarization of starlight is a result of the aligning of dust grains under the influence of an interstellar magnetic field.

with the composition of interstellar gas. By the late 1930s, Eddington had decided that the ionizing effects of ultraviolet radiation emitted by a star would wane rapidly with distance, so that most interstellar hydrogen would be in its neutral, or un-ionized form, known as HI. But in 1938 and 1939, Otto Struve and Christian T. Elvey, working with a specially designed nebular spectrograph on the eighty-two-inch telescope at the McDonald Observatory in Texas, detected a series of spectral lines, in various directions throughout the galactic plane, indicating the presence of extended regions of ionized hydrogen, or HII, which, according to Eddington, should not have existed.

THE STRÖMGREN SPHERES

The man who clarified the situation was a young Swedish-born astronomer named Bengt Strömgren, who had been publishing scientific papers since 1925, when he was scarcely seventeen. Strömgren's precocity was due in no small measure to the fact that his father was the director of the Copenhagen Observatory. Growing up on the observatory grounds, the younger Strömgren was, as he later wrote, "keenly interested in all that was going on in the place." He also developed an early interest in nuclear physics, fueled by proximity to Copenhagen University's Institute for Theoretical Physics, where Niels Bohr was helping to lay the groundwork of what would become known as quantum mechanics. The two interests came together in Strömgren's early studies, as he concentrated on the internal structure of stars. After spending

1957 Caltech scientists John Bolton *(below, left)* and Paul Wild suggested that an interstellar magnetic field would reveal itself through a characteristic splitting of the hydrogen spectral line.

1968 Using the sensitive radio dish at Green Bank, West Virginia, South African-born astronomer Gerrit Verschuur found the spectral-line split, proving that a magnetic field pervades interstellar space.

a year at the University of Chicago's Yerkes Observatory serving as an assistant to Otto Struve (just before Struve moved to Texas), Strömgren went back to Copenhagen in 1938.

A year after his return, and just before succeeding his father as director of the observatory, Strömgren published a landmark paper entitled "The Physical State of Interstellar Hydrogen." In it, he resolved the apparent conflict between Eddington's relatively severe limits on ionized hydrogen and Struve's observations of vast quantities of it. According to Strömgren's analysis, Eddington was essentially correct, except that "high-temperature stars, and especially clusters of such stars, are capable of ionizing interstellar hydrogen in regions large enough to be of importance in problems of interstellar space"—and large enough to account for Struve's findings.

Subsequent research has amended some of Strömgren's deductions. For example, he believed that the transition from ionized to neutral regions was quite abrupt. Astronomers now know, however, that HII regions (or Strömgren spheres, as they are also known) have sharp boundaries only in areas where interstellar hydrogen is relatively diffuse; in denser locations, where there is more opportunity for hydrogen ions to recombine with free electrons, the boundary is blurred. Similarly, Strömgren believed that ionized hydrogen was confined strictly to these spheres, but in fact there are very low levels of ionized hydrogen throughout interstellar space. Finally, Strömgren's assessment of the size of the HII zones—which he put at about 200 light-years in diameter—now appears to have been too high; the currently accepted figure is on the order of several tens of light-years, meaning that Eddington's estimate was more nearly on target. Despite these modifications, Strömgren's essential hypothesis remains intact, and his work marked a great advance in knowledge of the interstellar medium.

A SPIN-FLIP ATOM

With the outbreak of World War II and the Nazi occupation of western Europe, normal scientific pursuits all but halted. "During the war years," Strömgren later wrote, "I worked at the Copenhagen Observatory in relative isolation and with quite limited resources." The situation was much the same elsewhere, particularly in the occupied Netherlands, where the thriving Dutch astronomical community was virtually cut off from the lifeblood of science, the free flow of information. However, at the University of Leiden, Jan Hendrik Oort, the unofficial but universally recognized leader of the Dutch astronomers, found that the situation also presented an opportunity to ponder the data already available.

Some fifteen years earlier, Oort had refined and expanded Lindblad's theories on differential galactic rotation, and he continued to be interested in problems relating to the Milky Way's structure. He was thus intrigued when, in 1944, an American electronic engineer named Grote Reber published something never seen before: a map of the galaxy based on radio emanations from the sky. In 1937, Reber had set up a homemade radio dish in the backyard of his home just outside Chicago, and over the next dozen years he recorded radio signals from various parts of the celestial sphere. In 1944, he published an article describing his work and proposing that the emissions came from hot interstellar gas. Oort persuaded one of his graduate students, Hendrik van de Hulst, to review Reber's theories and observations. Then, almost as an afterthought, he suggested to van de Hulst that he look for spectral lines at radio wavelengths.

The idea held a particular attraction for Dutch astronomers. In the Netherlands, traditional optical astronomy is hampered by the country's often cloudy weather and low elevation, which puts it at the bottom of Earth's thick ocean of atmosphere. Short-wavelength radiation—from infrared through visible light and down to high-energy x-rays and gamma rays—is either blurred or blocked to varying degrees by atmospheric water vapor and other particles. Radio waves, in contrast, range from one centimeter to more than six miles in length, and anything under about sixty-five feet can easily penetrate this atmospheric blanket.

The twenty-six-year-old van de Hulst was hard at work on his doctoral thesis on the extinction of light by interstellar particles, but he turned promptly to the problem Oort had set out for him. At this point, observational data from the radio portion of the electromagnetic spectrum was scarce and sketchy. But if the ultimate goal was to map the galaxy, then he knew he should be looking at the distribution of interstellar hydrogen.

An atom of HI consists of a single electron orbiting a lone proton, each of which possesses the property of spin. When the spin of the electron is in the direction opposite that of the proton, the atom is in its lowest possible energy state *(page 53)*. Van de Hulst realized that a collision with another atom can cause the electron's spin to reverse, becoming parallel to the spin of the proton and slightly raising the energy state of the atom. The excited atom will usually undergo a second collision, which returns it to the lower energy state, flipping the spin of the electron back to its antiparallel direction. According to van de Hulst's calculations, this spin-flip would cause the atom to emit radiation at a wavelength of 21.2 centimeters (roughly the width of this book). Although the spin-flip transition was extremely uncommon for any individual atom, interstellar hydrogen clouds contain such an enormous number of atoms that, at any given moment, some of them would be in transition and emitting 21-centimeter radiation.

Van de Hulst believed that with the proper equipment and some concerted effort from the astronomical community, the 21-centimeter emission line could be detected. But he announced his theory in 1945, when the world was still at war. The search would have to wait.

THE BULLDOG AND THE GLOBULES

At about the time Oort diverted van de Hulst from his thesis on interstellar dust, Dutch-born astronomer Bart Jan Bok, a student at Leiden in the mid-1920s, turned his attention to it. In 1929, Bok had moved to Harvard, where he became a professor of astronomy known for, as one former student later put it, a "bulldog determination to get the answer, eventually, if necessary, by urging the invention and use of new techniques, and certainly using a lot of bright young men."

Traditionally, a lot of the "bright young men" in the Harvard astronomy department were actually bright young women, and in the mid-1940s, Bok joined one of them, Edith F. Reilly, on a study of roundish, dark objects that had long been observed against bright emission nebulae in various parts of the sky. Given recent speculation that stars were formed from condensations of interstellar dust and gas, the Harvard team noted in a paper published in 1947, they proposed to survey these "relatively small dark nebulae" on the grounds that they "probably represent the evolutionary stage just preceding the formation of a star."

Eliminating dark nebulae that appeared wispy and "wind-blown," Bok and Reilly defined their quarry as "approximately circular or oval dark objects of small size," which they labeled globules. After examining the best available photographs to identify regions where globules were visible against bright emission nebulae, the two astronomers found that the population associated with the bright nebula known as M8 was "unusually large." Of twenty-three possible candidates, sixteen were round and seven irregular.

Since the globules had to be between Earth and the bright nebula whose light they blocked, Bok and Reilly were able to estimate a maximum distance to the M8 globules of about 1,260 parsecs, or about 4,100 light-years. That information in turn allowed them to calculate the size of the globules themselves. Those in M8 ranged from 7,000 to 80,000 astronomical units in diameter, with twelve of the sixteen round ones measuring between 10,000 and 35,000 AU, or from 300 to 1,000 times the radius of the Solar System.

Although the smaller globules were completely opaque, some of the larger ones permitted enough background starlight to shine through to allow for more accurate estimates of their absorption rates. Based on these estimates, Bok later inferred the dust mass of the small globules to be about one-third the mass of the Sun. Assuming that dust made up only a small fraction of the objects' total mass, with the rest being gas, the Bok globules, as they came to be known, were very likely massive entities, perhaps in the process of gravitational collapse. Bok and Reilly urged their colleagues to carry out further studies of these protostars on the verge of birth.

A RADIO BREAKTHROUGH

Meanwhile, with the end of the war, Dutch astronomers turned with relish to the task of making radio astronomy a practical reality. The first task was to construct the equipment. Electronic hardware was in short supply in postwar Europe, so Oort and his colleagues at Leiden built their first receivers from scavenged German radar antennas. These early sets were ineffective, however, and it was not until 1951 that the Dutch were ready to make a serious assault on the 21-centimeter line.

The telltale radio signature of neutral atomic hydrogen is the result of random collisions that alter the direction in which the hydrogen atom's lone electron *(blue)* spins as it orbits the proton *(red)*. Normally, the electron and proton spin in opposite directions *(yellow and red arrows)*, and the atom remains at its lowest energy state. But when a collision reverses the electron's spin direction *(top)*, the parallel rotation of electron and proton raises the atom to a higher energy state. A second collision may then cause the electron to flip back to its original spin direction, at which point the acquired energy is released in the form of a photon of energy *(top, far right)* radiating at a wavelength of 21 centimeters.

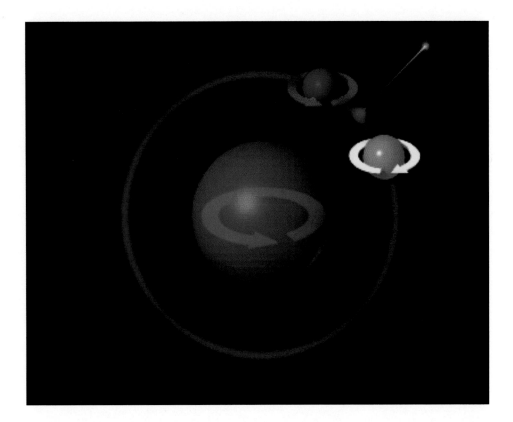

They might well have found it, but an untimely fire destroyed their receiver. While they regrouped, the honor fell to Harold Ewen, a graduate student working under Harvard physicist Edward Purcell. Like the Dutch, Ewen and Purcell had to cobble together their equipment. Their bulky horn antenna, mounted outside a window, was made of plywood covered with copper foil. Crude as it was, the antenna found its quarry: On March 25, 1951, the Harvard team recorded 21-centimeter radiation from neutral hydrogen in space.

As chance would have it, van de Hulst was visiting Harvard at the time. Ewen gave him a description of their antenna, which van de Hulst quickly relayed back to Leiden. Oort and engineer Alex Muller soon made the appropriate alterations in the rebuilt Dutch antenna, and on May 11, they confirmed the Harvard findings. In July, another team of radio astronomers in Australia made additional observations of the 21-centimeter line. All three groups jointly published their results in the journal *Nature* later that year.

Teams of observers in Australia and the Netherlands then began the arduous process of using the 21-centimeter line to map the distribution of HI gas throughout the galaxy, and by the late 1950s, most of the sky had been plotted in detail. The resulting Leiden-Sydney map *(page 57)* showed that neutral hydrogen tended to be concentrated in the galaxy's spiral arms. Overall, the gas formed an extremely flat, thin disk, which helped astronomers define the galactic equator.

INTERSTELLAR THERMOSTATS

While radio astronomers sketched the outlines of the galaxy in neutral hydrogen, Princeton astrophysicist Lyman Spitzer pondered the role of interstellar gas and dust in the formation of stars, a phenomenon that also seemed to be concentrated in the galaxy's spiral arms. In the mid-1930s, Spitzer had done graduate research under Eddington at Cambridge, where he also was impressed by, as he put it, the "physical lucidity" of Strömgren's work on the distinctions between HI and HII regions. In a series of papers published between 1948 and 1950, Spitzer reported on his investigation of interstellar clouds of gas and dust, which he believed were the birthplace of massive, luminous O and B stars. As a first step, he focused on how various methods of heating and cooling these clouds could yield the 10,000-degree temperature predicted by Eddington in 1926 for ionized HII regions, and hoped to resolve uncertainties over the temperature of neutral HI regions.

In later years, Spitzer would characterize his work as being "based on an ability to visualize a physical system and how it operates and thus to predict

The rosy Eagle nebula—made up primarily of ionized hydrogen (HII) irradiated by the ultraviolet light of massive young stars—provides a spectacular backdrop for the dark spots of dense gas and dust known as Bok globules *(indicated here by arrows)*. Evidently formed by pressure from ionized gas expanding at the outskirts of nebulae, these crowded plots of interstellar matter are nurseries for future stars.

its behavior without benefit of mathematics—though mathematics provides a much needed check." Accordingly, in tackling this project, he envisioned, and then calculated, the behavior of electrons, ions, neutral atoms, and dust grains in interstellar space. Spitzer's computations confirmed the processes predicted by Eddington for HII regions, namely, that the ionization of neutral hydrogen atoms by ultraviolet radiation from nearby stars maintained a temperature of about 10,000 degrees, despite cooling that involved collisions between electrons and ions.

For HI regions, Spitzer had to devise new heating and cooling mechanisms. In a paper published in 1950, Spitzer and Princeton doctoral candidate Malcolm P. Savedoff proposed several methods of cooling: Collisions between hydrogen atoms and dust grains would lead to a net loss of energy, as would collisions between electrons and ions. A third method involved the excitation of electrons in ions of elements that are heavier than hydrogen—carbon, silicon, and iron in the case of the HI zones—a process that would tend to radiate away excess energy.

The two scientists also presented two mechanisms for the heating of HI clouds, depending on the cloud's density. In dense clouds, the primary source of heat was the capture and reemission of electrons by ions of heavy elements. At low densities, colliding with particles known as cosmic rays, which move at near light-speed, was the more significant factor. Both these sources of heat are relatively weak, Spitzer and Savedoff noted, especially in relation to the three efficient cooling mechanisms at work in HI regions. As a result, temperatures would hover between 30 and 100 degrees Kelvin, with an average of about 60 degrees.

Although later studies show that heating by cosmic rays is less intense than Spitzer believed, his identification of their integral role in the dynamics of the interstellar medium was greeted with general, almost embarrassed approval by his colleagues. Cosmic rays are an astrophysical mystery dating from the late 1700s, when French physicist Charles Augustin Coulomb noticed that the electric charge on an object suspended by a silk thread somehow leaked away into the air. With the discovery of radioactivity in 1896, scientists speculated that charged atoms, or ions, produced by radioactive material in Earth's soil might somehow be drawing off the charge. They were wrong, but proving it would take more than thirty years.

RAYS FROM SPACE
Among the first to test the proposition was a young and daring Austrian physicist named Victor Franz Hess, who made ten balloon ascents in 1911 and 1912, the loftiest of which reached an altitude of about three miles. If the theory was correct, the intensity of ionizing radiation would decrease with altitude. Instead, just the opposite occurred. By making ascents at night and even during a solar eclipse, Hess ruled out the possibility that the ionizing rays came from the Sun, but the origin of this "Hohenstrahlung," or "radiation of the heights," remained unknown.

Then, in 1925, Robert Millikan, a Nobel prize-winning physicist at the California Institute of Technology, performed his own series of experiments. Millikan and one of his students, G. Harvey Cameron, trekked to two high-altitude lakes in California. Readings taken six feet below the surface of 12,000-foot-high Muir Lake matched those taken at the surface of Lake Arrowhead, at an altitude of only 5,000 feet. Since the absorbing power of six feet of water balanced that of the 7,000 feet of air between the surfaces of the two lakes, the experiment proved that the radiation did not originate in Earth's atmosphere, which instead functioned, in Millikan's words, as an "absorbing blanket."

Because the radiation did not have a specific source but seemed to come from everywhere in the universe, Millikan dubbed the radiation "cosmic rays." (For a while the press referred to them as "Millikan rays," which did not sit well with Hess, who was miffed by all the attention lavished on Millikan. However, Hess eventually did receive a Nobel prize of his own for the work.) In the 1930s, a fellow Nobel laureate at Chicago, Arthur Holly Compton, showed that cosmic rays are not very short wavelength electromagnetic radiation, as their power of penetration had suggested, but are instead charged particles (protons, for the most part) that carry enormous amounts of kinetic energy. Some of the most energetic hit Earth with as much punch as a powerfully served tennis ball—even though a tennis ball is many trillions of times more massive.

POLARIZED STARLIGHT

So much energy coursing through the galaxy was bound to have an effect on the interstellar medium—as Spitzer and Savedoff demonstrated with their work linking cosmic rays to the heating of hydrogen clouds—but astrophysicists have yet to pinpoint the precise origin of the cosmic particles. (Possible candidates abound, including supernovae, x-ray binaries, and stellar winds from hot stars.) Although recent observations have shown that cosmic rays have roughly the same composition as that of the Sun and other stars—about 87 percent hydrogen, 12 percent helium, and the rest divided among heavier elements—the particles also show an anomalous abundance of otherwise rare elements such as beryllium and lithium, which occur as cosmic rays up to one million times more often than they should.

Whatever their source, the extremely high energies displayed by cosmic rays lead astronomers to believe that they must receive an additional boost to replace a natural energy loss on their lengthy voyage to Earth. In 1949, the celebrated Italian-American physicist Enrico Fermi suggested that cosmic rays pick up extra energy by way of interactions with interstellar magnetic fields. In the Fermi mechanism, as it is known, moving clouds of ionized hydrogen carry magnetic fields along with them; when a cosmic-ray particle collides with a cloud, it spirals along and reflects off the lines of magnetism within the cloud and thus picks up fresh energy. (Although the interstellar magnetic field is thought to have a part in accelerating cosmic rays, astrono-

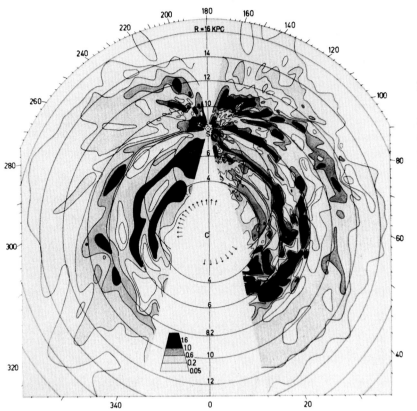

This chart of HI emissions in the Milky Way galaxy—known as the Leiden-Sydney map for the sites at which the observations were compiled—reveals high HI concentrations *(dark shading)* in circular bands corresponding to the galaxy's spiral arms. The Sun *(S)* appears here as a dot directly above the galactic core *(C)*; the uncharted cone below the Sun represents that portion of the galaxy that is rotating parallel to Earth and whose emissions thus cannot be distinguished by Doppler shift.

mers do not now consider it the chief agent.)

As it happened, a team of scientists working at the McDonald Obseratory had found evidence of interstellar magnetism just the previous year. William Hiltner and John Hall were attempting to confirm a prediction, made three years earlier by the noted Indian-American astrophysicist Subrahmanyan Chandrasekhar at the University of Chicago, that in some cases photons from the interior of a star will reflect off atmospheric electrons so that the star's light is polarized. That is, instead of radiating in all planes, the observed light consists of electromagnetic waves preferentially oriented in one plane.

Since individual stars, except for the Sun, are too far away to permit observations of a stellar sphere, the two astronomers concentrated on eclipsing binary stars. As one star passed in front of the other, the polarization in the atmosphere of the background star would be revealed, or so they thought.

Hiltner and Hall employed two light-gathering systems: a simple Polaroid filter hooked to a telescope with a photoelectric sensor, and a rotating prism attached to a telescope, also with a photoelectric sensor. Both devices assured that only polarized light would be collected. To their considerable surprise, the astronomers found polarization stronger than they had expected—but it was not a product of stellar atmospheres.

First, they had expected that the light would be polarized parallel to the plane of the star's orbit. Instead, the plane of polarization was aligned with the plane of the galaxy. Furthermore, if stellar atmospheres were responsible for polarizing starlight, the effect should be detected in all regions of the sky. Instead, Hiltner and Hall could find no polarization in binaries that were far from the galactic plane. Finally, the researchers found that the degree of polarization was directly related to the amount of reddening of the starlight by interstellar dust.

Later that year Hendrik van de Hulst addressed all of these effects. Three years earlier, van de Hulst had finished his dissertation—a theoretical examination of the absorption and scattering of light by particles of varying composition and size—and had then gone on to do further work on the subject at Yerkes Observatory. Astronomers knew in a general way that the extinction of starlight by interstellar dust is wavelength dependent. That is, light at the shorter-wavelength, or blue, end of the visible spectrum is blotted out more efficiently than longer-wavelength red light. However, as van de Hulst pointed out in the first of two papers published in 1949, very little in the way of

In the Glow of Ionized Hydrogen

Like a tuft of pink cotton candy in the blackness of space, a so-called emission nebula known as the Lagoon nebula *(right)* spreads its brilliant glow across ten light-years and holds within it a cluster of hot young O- and B-type stars. Emission nebulae are vast concentrations of interstellar gas that glow because of a process called photoionization. Gas in the interstellar medium is primarily neutral atomic hydrogen, or HI. But when condensed into clouds and bombarded with the ultraviolet radiation given off by hot young stars, neutral hydrogen becomes ionized. The UV radiation strips an electron from a hydrogen atom, leaving a charged proton, or ion. When the free electron is eventually captured by another hydrogen ion, it recombines to form a hydrogen atom, releasing a photon of long-wavelength red light—and creating an emission nebula's characteristic reddish glow.

detailed analysis of the problem had been carried out. He therefore proceeded to calculate so-called extinction curves produced by particles assigned certain properties such as size, shape, charge, and magnetism, and found that bumps in the curve at particular wavelengths could yield information about the nature of the grains responsible for the extinction.

Upon comparing the observed extinction of starlight with his theoretical computations, van de Hulst concluded that interstellar dust grains were primarily particles about .0004 centimeter in diameter. He further concluded that the particles consisted of ice, formed by condensation, as suggested fourteen years earlier by Lindblad. The process of condensation, however, required initial "nucleation sites"—very tiny core particles around which the ice condensed—whose origins remained uncertain.

Although van de Hulst's calculations in this paper assumed that the particles were spherical and homogenous in composition, his second report proposed that the polarization of starlight discovered by Hiltner and Hall was produced not by spherical particles but by elongated ones. Noting that light striking spherical particles would be absorbed or scattered in every direction, rather than preferentially in just one plane, he suggested that particles shaped like grains of rice, for instance, could polarize light if they were all aligned in a consistent direction.

THE CASE FOR ALIGNMENT

In 1951, Caltech scientists Leverett Davis and Jesse Greenstein confirmed van de Hulst's speculations. Greenstein had written his 1937 doctoral thesis on the subject of the interstellar medium but had subsequently taken up stellar

The photoionization process that creates an emission nebula's red color begins when a photon of high-energy, short-wavelength ultraviolet radiation from a hot star strikes a neutral hydrogen atom and knocks away an electron *(left, top)*, yielding a negatively charged free electron and a positively charged hydrogen ion. When the electron later collides with an ion, it may be captured in any one of several orbits, or shells, around the nucleus, such as shell 3 *(middle)*. Within a fraction of a second, the electron will jump to shell 2 *(bottom)*, releasing energy in the form of a photon of long-wavelength red light. Since hydrogen's strongest emission line at optical wavelengths lies in the red portion of the spectrum, the nebula is visible from Earth as a brilliantly fluorescing red cloud.

spectroscopy until Hiltner and Hall's polarization studies prompted him to revisit the earlier topic. Together with Davis, he published a long, closely reasoned, highly mathematical paper in the September issue of the *Astrophysical Journal,* examining all the mechanisms that could cause interstellar dust grains to align.

Following generally accepted notions regarding the relative abundance of elements in space—namely, that they match those of the Sun and other stars—Davis and Greenstein assumed that the dust particles were some sort of crystalline substance, such as ice or ammonia, "with various heavier atoms present as diffuse impurities." In particular, they wrote, "about one atom in one hundred should be iron or a similar atom having a large magnetic moment." Random collisions with gas atoms and other particles would cause the interstellar grains to spin. In the absence of a magnetic field, the two scientists reasoned, the spinning grains would be randomly oriented, producing no polarization of light. But an interstellar magnetic field would exert an aligning torque on the grains, orienting them with their long axes perpendicular to the field so that they would spin like an airplane propeller around the shaft of the magnetic field.

Such an orientation would account for the observed polarization of starlight, since only light waves that are traveling parallel to the magnetic field would get through the barrier of aligned grains. "Over regions of several hundred parsecs in the Milky Way," Davis and Greenstein hypothesized, "the magnetic field is mainly parallel to the plane of the galaxy, perhaps nearly uniform along a spiral arm or perhaps making random whirls mainly in the plane of the galaxy." The astronomers estimated a strength for the field of just

WHEN LIGHT MEETS DUST

Interstellar dust, unlike interstellar gas, does not announce its presence by producing comparatively narrow absorption lines in the optical spectra of stars *(pages 24-25)*, largely because solids like dust grains absorb a broad range of wavelengths. Yet the accumulation of this dust along the line of sight from Earth has noticeable effects on starlight. Dust can form so-called dark nebulae—light-blocking concentrations that are detectable only when viewed against a bright background, just as clouds passing in front of the full Moon look black to an observer on the ground. At other times, clouds of interstellar dust glow with reflected starlight, much as their sunstruck earthly counterparts turn rosy at sunrise and sunset.

Another effect of dust on light is known as polarization. The light's electric field, instead of radiating in all planes, is preferentially oriented in one plane. Light passing through interstellar dust is polarized for two reasons. First, the dust grains are elongated rather than spherical; round grains would absorb and scatter light in all directions. Second, the grains are aligned in one direction by the interstellar magnetic field; without the field, elongated grains would be randomly oriented and there would be no polarization.

Visible light, like all other forms of electromagnetic radiation, consists of periodically varying electric and magnetic fields that vibrate perpendicularly to each other and to the direction of travel. As shown here, light waves with electric fields that happen to be oscillating in the plane of the long axis of interstellar dust grains will be largely blocked *(left)*, but waves whose fields oscillate perpendicularly to that axis will pass through more easily *(below)*.

The absorption and scattering of light by interstellar dust is wavelength dependent. Short-wavelength blue light, for example, is more likely to be scattered than longer-wavelength red light. As a result, reflection nebulae, which shine with light reflected from stars off to one side of the line of sight from Earth, have a somewhat bluish cast *(below)*. But if stars are viewed directly through a region rich in dust, the scattering of blue light leaves mostly red light to penetrate the cloud *(below, right)*. In a sense, the light is not really reddened but "de-blued."

Dust in a reflection nebula in Orion scatters short-wavelength

A dusty cloud, Barnard 86, obscures background stars and

10 to 100 microgauss (one-millionth of a gauss, which is the standard unit of measure for magnetic field strength), or about one ten-thousandth the strength of Earth's own magnetic field. Two years later, Enrico Fermi and Subrahmanyan Chandrasekhar suggested a calculation of about 7 microgauss. The value remains controversial today, although modern estimates range from 2 to 5 microgauss in the diffuse interstellar medium, with higher values in interstellar clouds.

SPLITTING THE HYDROGEN LINE

Four more years passed, and then in 1957, Caltech scientists John Bolton and Paul Wild proposed a method by which such a measurement might finally be achieved. The researchers suggested that a very sensitive radio telescope might be capable of measuring the weak interstellar magnetic field by observing its influence on the 21-centimeter line of neutral hydrogen in a phenomenon known as the Zeeman effect. In 1897, Pieter Zeeman, yet another trailblazing Dutch scientist at the University of Leiden, had discovered in a series of laboratory experiments that in the presence of a magnetic field, spectral lines broadened.

Later investigators learned that rather than simply broadening, such lines actually split into two or more distinct lines. In the case of hydrogen, the magnetic field influences the spin-flip transition responsible for the 21-centimeter line by speeding up or slowing down the spin of the electron. This slight alteration of energy levels divides the original spectral line into two observable components of slightly different energies, or frequencies, and polarizations. The stronger the field, the wider the separation between frequencies and therefore between lines.

Bolton and Wild had posed a tremendous challenge for radio astronomers. Zeeman splitting had been observed in the optical spectra of stars with extremely powerful magnetic fields on the order of 5,000 gauss, or 10,000 times stronger than Earth's. It would be far more difficult to detect in a galactic field four orders of magnitude weaker than Earth's. Moreover, the wavelength separation would be just one-thousandth the width of the narrowest 21-centimeter emission line—like distinguishing a single thread in a yard-wide bolt of cloth from ten feet away.

When the Zeeman splitting was finally found, it was, appropriately, a scientist of Dutch ancestry who made the discovery. By the late 1960s, South-African-born Gerrit Verschuur—like many of his colleagues at various major observatories—had spent hundreds of hours searching fruitlessly for the Zeeman effect, only to be frustrated by the inadequacy of his equipment. Early radio telescopes tended to emit so much internal noise that the faint signals coming in from space frequently were obscured.

The thirty-year-old Verschuur, after seven years at Jodrell Bank radio observatory in England, had all but given up by 1967, when he arrived as an assistant scientist at the National Radio Astronomy Observatory in Green Bank, West Virginia. But the 140-foot Green Bank dish was a state-of-the-art instrument, so quiet that Verschuur, with some trepidation, once again took up the quest for what had become, as he later wrote, radio astronomy's "version of the Holy Grail."

In mid-1968, following a ten-day observational run in May, he was processing the data by hand because the necessary computer programs were not yet available. "For weeks," he later wrote, "I did very little other than that: add, divide, and occasionally eat and sleep." Even when the computer programs were finally ready to use in early July, Verschuur still had to plot the resulting graphs manually. As he was working on data from a dense gas cloud in the Perseus arm of the Milky Way galaxy, he was suddenly elated at the slightly offset pair of lines forming on the page before him: "A huge Zeeman effect!" Later that year, his elation was even greater when, rather than spotting the effect after the fact, he actually observed it in real time, as the data from the radio telescope was translated into a plot on an oscilloscope. Shouting "It's there! It's real," Verschuur danced for joy. His subsequent measurements put the strength of the magnetic fields in the cloud at between ten and twenty microgauss.

The finding was the culmination of decades of scientific detective work, the piecing together of clues that seemed at first to have no connection but which finally yielded up a roster of hidden players: first gas; then dust and cosmic rays; and then, like a kind of cosmic scaffolding, the galactic magnetic fields. Once relegated to the scientific back burner because it was believed to be bland and peaceful, the darkness between the Sun and its billions of cousins in the galaxy was proving to be a region of unexpected variation and dynamism whose secrets were only beginning to unfold.

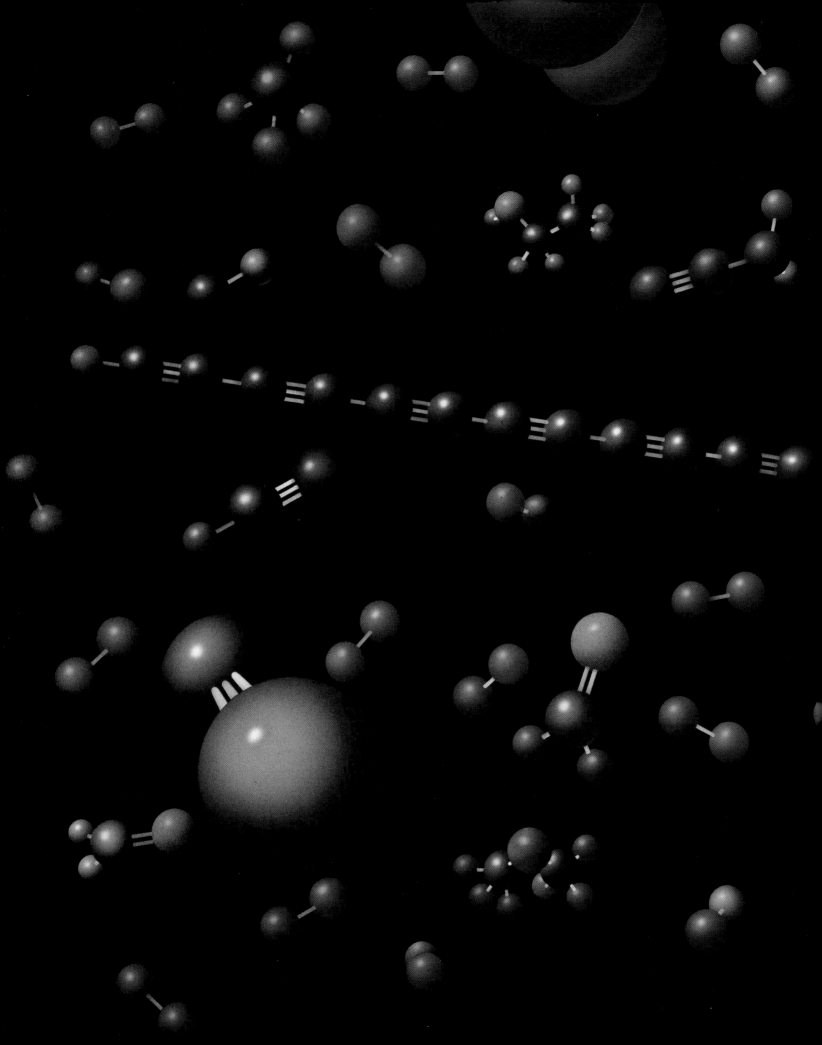

Scattered through the vast darkness between stars, the molecules of interstellar space range from very simple to incredibly complex. Formed when two or more atoms collide and chemically bond, these molecules of hydrogen, carbon monoxide, and scores of other compounds generally make up a tenuous soup—a trillion trillion times less dense than stars or planets. Yet under certain conditions, far-flung molecules may collect into clouds that serve as seedbeds for new stars. These stellar breeding grounds are yielding their secrets, thanks to the tendency of molecules to absorb or surrender specific amounts of energy as they collide with one another or with other stimuli such as cosmic rays. The quantum of energy that each type of molecule exchanges in the process shows up at a precise wavelength along the electromagnetic spectrum. By analyzing these spectral signatures, astronomers can not only classify interstellar molecules and gauge their relative densities but also draw certain inferences about how complex molecules evolve from simpler forms in deep space.

The essential building block of molecular clouds is hydrogen, which makes up more than 50 percent of the interstellar mass. Commonly produced when two hydrogen atoms *(blue)* bond around a speck of dust *(pages 68-69)*, molecular hydrogen may then be ionized by cosmic rays so that it becomes receptive to other atoms in the interstellar mix, such as oxygen *(green)*, carbon *(black)*, nitrogen *(red)*, and sulfur *(yellow)*. The resulting compounds include esoteric molecules that are nonexistent on Earth, and whose intricate geometry can only be guessed at by astrochemists. Thus far, 92 molecules have been identified in the interstellar medium. Some astronomers believe that as many as 500 molecular species inhabit that realm—a remarkable reversal of the long-held view that the harsh environment of deep space would prevent the formation of anything more complicated than a simple, two-atom molecule.

PROFILE OF THE MOLECULAR BUILDING BLOCK

To trace the evolution of complex molecules in space, astronomers first had to assay the simplest of compounds: diatomic hydrogen. The prevalence of atomic hydrogen between the stars, clearly signaled by 21-centimeter radio emissions, suggested that such atoms might well come together often enough to create a vast store of molecular hydrogen that could provide the basis for other interstellar compounds. But diatomic hydrogen proved difficult to pin down. Typically, molecules display their spectral signatures when a collision increases their rate of rotation and the acquired energy is then released in the form of a photon *(below)*. The release of radiation is made possible by the asymmetry of most molecules; in such cases, the collision transforms the molecule into a kind of radio transmitter, propelling the molecule's

Following a collision, a carbon monoxide molecule rotates energetically in the interstellar medium *(below, left)*. In time, the molecule relaxes to a slower rate of rotation, emitting energy in the form of a photon that exactly equals the kinetic difference between the two rates of spin *(below, right)*. The capacity of the rotating carbon monoxide molecule to generate a photon is based on its asymmetry, as detailed opposite.

center of electromagnetic charge around its center of mass *(below, right)* until a photon is emitted—usually at short radio wavelengths—and the molecule reverts to its ground state. Because diatomic hydrogen is symmetrical, however, its centers of charge and mass occupy the same place, and it produces no such signal through rotation.

Faced with the radio silence of molecular hydrogen, astronomers sought and detected the signature of carbon monoxide, an asymmetrical molecule that is energized by collisions with hydrogen molecules. Analysis of that signature reveals a ratio of hydrogen to carbon monoxide of about 10,000 to one. Traces of carbon monoxide pervade huge areas of space, confirming the primacy of molecular hydrogen in the interstellar medium.

In spinning but symmetrical molecules, such as diatomic hydrogen *(top),* the center of electromagnetic charge and the center of mass share the same point *(purple and red),* and no signal is generated. In asymmetrical molecules consisting of different types of atoms, such as carbon monoxide *(bottom),* the centers of charge *(purple)* and mass *(red)* diverge. As the center of charge whirls around the center of mass, energy is emitted.

An Atomic Meeting Ground

Once astronomers could gauge the abundance of interstellar molecular hydrogen, they could evaluate various theories as to how the molecules evolved. The simplest way for hydrogen molecules to form is through collisions of hydrogen atoms. Known as radiative association, this process has evidently been occurring since the earliest era of the universe. But only about one collision in ten billion actually results in a molecule. Even if the Milky Way was always as crowded with hydrogen atoms as it appears today, that rate is still far too low to account for the high incidence of diatomic hydrogen in the galaxy's clouds.

Plainly, some other mechanism is at work as well, one involving an agent that promotes molecular bond-

The formation of an interstellar hydrogen molecule begins when a grain of dust that has been hurled into deep space by a dying star is struck by a hydrogen atom. As the dust particle absorbs the energy of the impact, the atom sticks to its surface.

After a long interval, another hydrogen atom strikes the dust particle and adheres. While such encounters are random and rare, there is enough hydrogen in interstellar space—about one atom per cubic centimeter—to cause dual impacts on many dust particles.

ing. The obliging intermediary, it turns out, is dust: microscopic, elongated particles of carbon, silicon, and other elements ejected by dying stars. When struck by a hydrogen atom, a dust grain absorbs the kinetic energy, allowing the atom to stick to its surface. Once a second hydrogen atom adheres to the same grain, the stage is set for the formation of a molecule. Although dust makes up only about one percent of the interstellar mass, it brings atoms together with remarkable efficiency, producing thousands of molecules for every one generated by radiative association. The unions that result are long-lasting, allowing vast reserves of diatomic hydrogen to build up into molecular clouds.

Once the two hydrogen atoms are attached to the grain of dust, they find each other and form a chemical bond, yielding a molecule whose energy is less than that brought to the union by the two atoms. The excess energy is absorbed by the dust particle.

In the aftermath of the bonding process, the energy that is absorbed by the dust particle makes it so warm that the molecule evaporates from the surface and drifts off into space. Typically, the formation of a hydrogen molecule takes about 100,000 years.

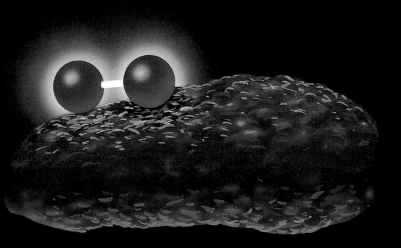

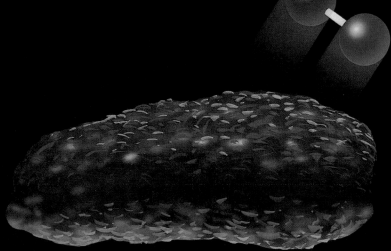

Forging a Chain

The formation of simple hydrogen molecules around dust particles gives substance to deep space, but an infusion of energy is required to account for the development of such diverse compounds as water, ammonia, and methane. Since the temperature and energy levels in interstellar space are very low, the journey from diatomic hydrogen to more complex forms must be fueled by an outside source. That energy comes from cosmic rays: charged nuclear particles traveling at near the speed of light.

When a cosmic ray hits a simple hydrogen molecule, it triggers a reaction that leads to the vital link in the interstellar molecular chain—the hydrogen ion known as $H3+$, which contains three protons and two elec-

The sequence of events leading to complex interstellar molecules begins when a simple hydrogen molecule *(below, left)* is hit by a cosmic ray, which knocks a negatively charged electron from the molecule *(below, center)*. The impact reduces the neutrally charged hydrogen molecule to a positively charged ion known as $H2+$, containing two protons and one electron. $H2+$, in turn, is soon transformed by an encounter with a neutral hydrogen molecule *(below, right)*.

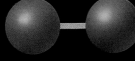

trons. This positively charged compound readily reacts with other molecules or atoms, initiating long sequences of combinations and disassociations that result in the formation of complex molecules. Like billiard shots on a huge cosmic table, some of these sequences may involve more than 1,500 reactions, presenting a fiendishly difficult problem for earthbound chemists who are trying to duplicate the reactions in their laboratories. But by understanding the many steps involved in the formation of a known molecule, scientists can predict other possible permutations— and perhaps account for some of the mysterious spectral signatures that cannot be attributed to molecules found on Earth.

The reaction of H2+ with a neutral hydrogen molecule yields an H3+ ion *(below)*, containing three protons and two electrons (a neutral hydrogen atom is released as a by-product). The H3+ ion is endlessly adaptable, reacting with other components of the interstellar mix in sundry ways.

Shown below are three of the more common molecular products of chain reactions instigated by H3+: ammonia (three hydrogen atoms and one nitrogen atom), methane (four hydrogen atoms and one carbon atom), and water (two hydrogen atoms and one oxygen atom). Even a fairly simple product such as water requires several steps beyond the initial reaction of H3+ with oxygen; the synthesis of more elaborate molecules is a hundred times more complicated.

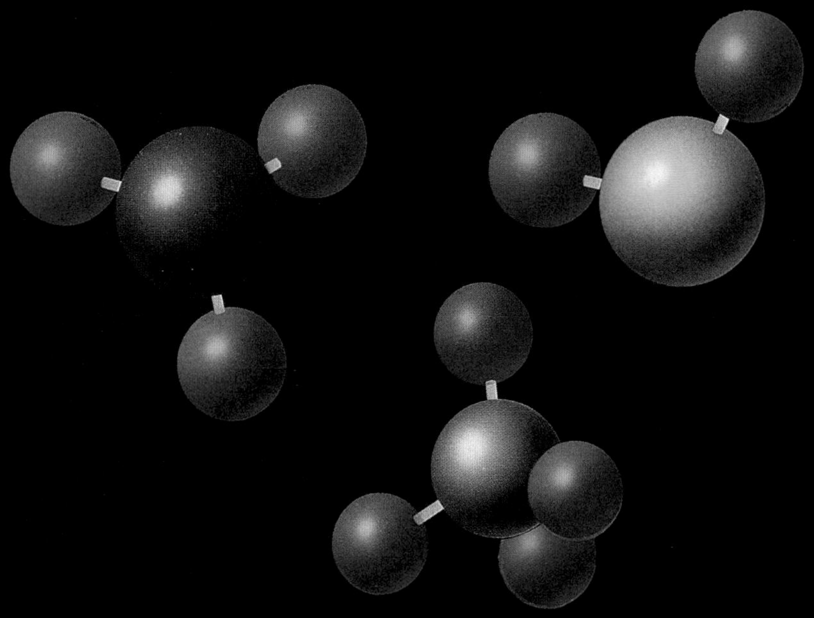

THE OUTER REACHES OF COMPLEXITY

Larger, more elaborate molecules produce distinct signatures in the infrared portion of the electromagnetic spectrum. But spectral analysis in that zone is complicated by solids such as dust grains that also produce infrared emissions. Astrochemists consistently find fifteen to twenty signatures at infrared wavelengths that cannot be positively identified. A few have been attributed to known molecules whose signals are being confused by interactions with the very dust grains that helped create them. One notable case involves intriguing infrared emissions coming from vast clouds that exist just above and below the plane of the Milky Way. (These regions may be generating as much as 40 percent of the infrared radiation in the galaxy.) After careful analysis, astronomers have ascribed the signals to a complex class of organic ring molecules known as polycyclic aromatic hydrocarbons (PAHs). If this attribution is correct, it raises

The structural basis of the complex PAH molecules that may exist in the interstellar medium is the benzene molecule *(below, left)*. The alternating single and double bonds between the carbon atoms in the hexagonal ring cause the molecule to be very stable.

In an elegant extension of benzene's hexagonal form, six rings with shared walls surround the hub of the coronene molecule. Composed of twenty-four carbon atoms, coronene is one of the PAHs whose presence in deep space has been inferred from spectral signatures.

interstellar chemistry to a new level of sophistication.

Astrochemists have also posited the existence of molecules that would fill an important niche in the interstellar environment but have yet to be linked to specific signatures. These speculative compounds include conglomerations of carbon atoms known as fullerenes, for the inventor of the geodesic dome, R. Buckminster Fuller. Such molecules, which incorporate the hexagonal ring structure of PAHs, contain scores of carbon atoms locked tightly together in a form that resembles one of Fuller's sturdy domes. So stable is the design that a leading proponent of the idea believes fullerenes may have been the first objects with surfaces to form after the Big Bang. Among the virtues of this hypothesis is that the compounds would be tough enough to survive impact with other molecules, thus helping to account for the formation of dust, comets, asteroids, and planets.

Incorporating features of benzene—including the hexagonal ring and the alternating single and double bonds—this symmetrical fullerene contains sixty carbon atoms. Its spherical configuration may make it the most stable molecule in the interstellar medium.

Fullerenes may begin taking shape when a few carbon atoms bond together, creating a sheetlike network with hexagonal edges. Ultimately, the sheet folds over on itself, and the edges meet. Although the optimum result is a rounded fullerene, less symmetrical forms might also result *(below)*.

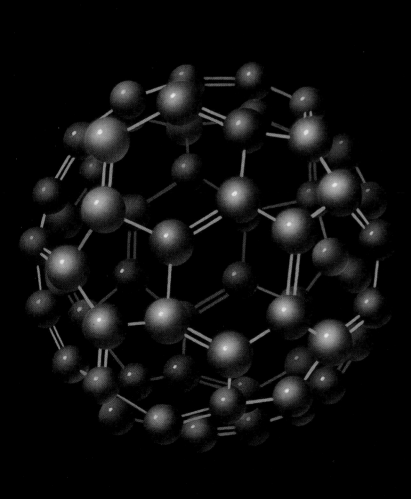

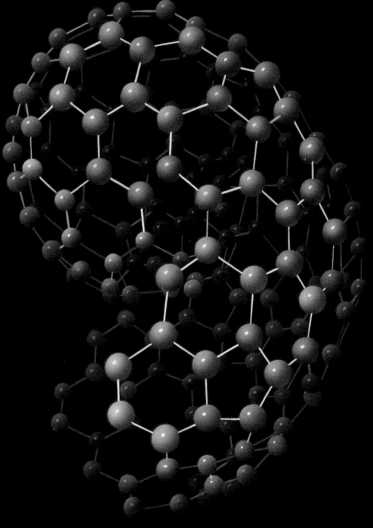

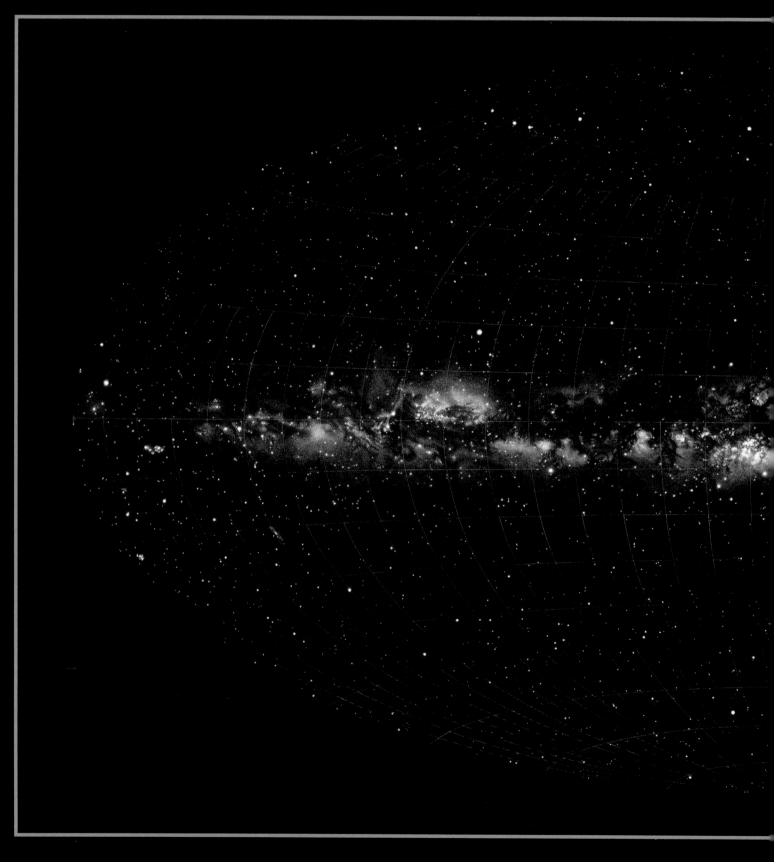

OF STARS

Dark patches split the luminous central plane of the Milky Way galaxy in a composite view produced by projecting the details of a sweeping photographic survey of the sky onto an oval grid. Astronomers have determined that this so-called Great Rift and others like it are vast clouds of molecular gas that can give birth to new stars.

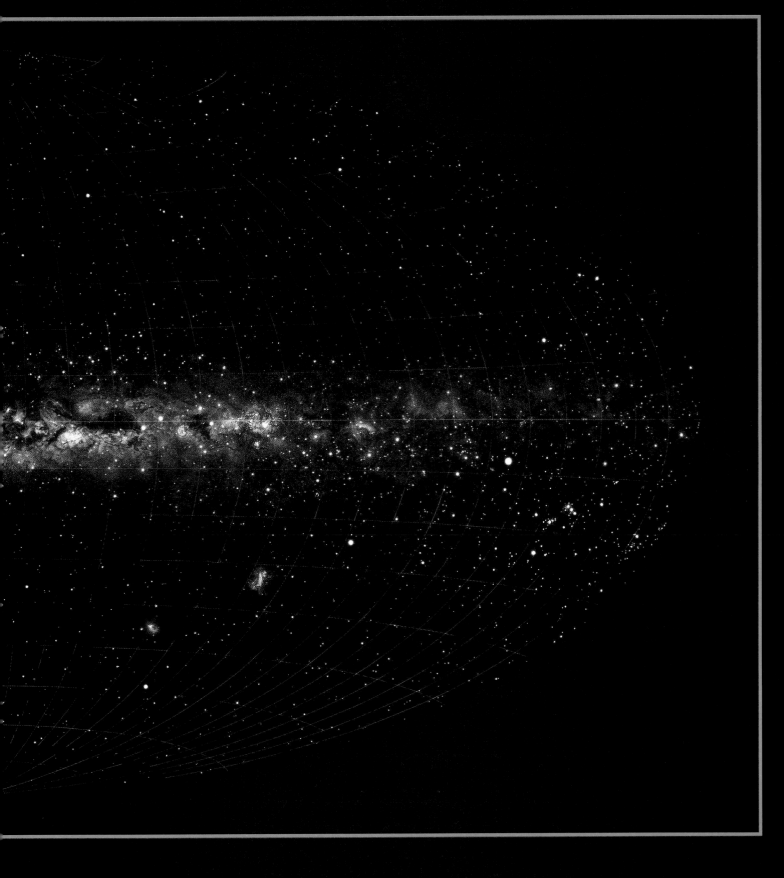

any centuries ago, the Aboriginal tribespeople of Australia took note of a familiar likeness in the shimmering vault of their midsummer night's sky. Against the silvery band that would come to be known as the Milky Way, they perceived the huge silhouette of an emu, the flightless, ostrichlike creature of their homeland. In photographs of the Milky Way, the Emu's head and beak are formed by a dark region that modern astronomers call the Coalsack, and its long neck is a narrow band running through the constellation Centaurus to the constellation Norma. Dark patches in Scorpius and Sagittarius form the Emu's body, and its long, thin legs are strips found in Ophiuchus.

The Emu is made up of several large complexes of gas and dust clouds that obscure the light of stars lying behind them. Thousands of similar aggregations are strung along the spiral arms of the Milky Way, and together such clouds may account for as much as seven percent of the total mass of the galaxy. Most are relatively small, ranging from three to a thousand times the radius of the Solar System in diameter and containing a mass of dust and gas equal to perhaps several hundred solar masses; these are the rounded objects known as Bok globules, which give birth to low-mass stars. Perhaps 5,000 clouds are colossal, on the order of 70 to 250 light-years across, with masses between a hundred thousand and five million times the mass of the Sun. These shadowy behemoths, the largest known objects in the Milky Way, are the stellar nurseries of the galaxy, birthplaces of the brightest, most massive, and shortest-lived O- and B-type stars.

In recent years, studies of the dark clouds have yielded a wealth of discoveries. Scientists have learned that the gas in the clouds is made up not simply of atoms, as long held, but of molecules, the next step up in the chain of matter. Although the gas is primarily molecular hydrogen (H_2), it also contains significant amounts of more complex forms; nearly a hundred molecules have been identified so far, including alcohol (C_2H_5OH)—although a volume of space the size of the planet Jupiter would be required to collect enough to fill a shot glass. Perhaps the most intriguing discovery of all is that a sizable fraction of the complex molecules in molecular clouds are organic, carbon-based compounds—the basis of all terrestrial life.

Although Arthur Eddington and others had speculated about the existence of interstellar molecules as early as the mid-1920s, the notion was essentially dismissed for several decades because no one could explain how individual atoms would be able to join up in the vastness of space. According to the reasoning of the day, the few molecules that might form there would be almost instantly torn apart by high-energy ultraviolet rays from hot stars and cosmic rays from nearby supernovae.

MOLECULAR EVIDENCE

As is often the case, however, observation confounded theory. In 1936, Walter Adams, director of California's Mount Wilson Observatory, was working with Theodore Dunham on spectrographic studies of certain bright, young O and B stars using Mount Wilson's 100-inch telescope. In examining a series of spectrograms, the pair found several narrow absorption lines at wavelengths where such lines had not been detected in stellar spectra before. The sixty-year-old Adams, who had helped design both the 100-inch instrument at Mount Wilson and the 200-inch Hale Telescope just being built for Palomar Observatory, was highly regarded by his colleagues both as an observer and as a spectrographer. After some analysis, he and Dunham succeeded in identifying four of the lines as having been produced by ionized sodium, neutral potassium, neutral calcium, and ionized titanium—all of which were new discoveries in interstellar space.

But the two astronomers were unable to name the sources of the remaining lines, including one at 4,300.3 angstroms and another at 3,874.6 angstroms. In the following year, another team of researchers theorized that the line at 4,300.3 angstroms might be the signature of methylidyne, a diatomic, or two-atom, molecule of carbon and hydrogen, two elements known to be abundant in space. But scientists had no way to confirm this hypothesis, and the issue remained unresolved for three years.

Then, in 1940, astrochemist Andrew McKellar of Canada's Dominion Astrophysical Observatory in Victoria, British Columbia, carried out laboratory experiments showing that detectable lines in the violet and near-ultraviolet end of the visible spectrum would occur when molecules underwent certain low-energy transitions. McKellar provided tentative identifications for the sources of two lines found by Dunham and Adams—methylidyne at 4,300.3 angstroms and cyanogen (CN) at 3,874.6 angstroms—and also predicted a line at 3,934.5 angstroms for sodium hydride.

Within a few years, scientists found absorption lines for methylidyne and the methylidyne ion (CH+) in the spectra of many more bright, blue-white stars. But even the growing evidence for the omnipresence of these molecular species did little to counteract the prevailing belief that molecules were of virtually no significance in space. The examples so far were all simple diatomic molecules, were found only in trace amounts, and were thought to be short-lived. More than two decades would elapse before another molecular species was discovered.

As it happened, not all scientists dismissed the possibility of interstellar molecules out of hand. The 1951 detection of neutral atomic hydrogen via the 21-centimeter line had led physicist Charles Townes and his colleagues at Columbia University to consider looking for molecules at radio frequencies rather than visible wavelengths. Before such an undertaking could begin, however, they would need precise measurements of the so-called rest frequencies transmitted by molecules in the laboratory.

Townes, who was born and raised in Greenville, South Carolina, was the right man for the job. He attended his hometown's Furman University, where he attained degrees in both modern languages and physics in 1935. After

INTERSTELLAR HUNTERS

The hunt for interstellar molecules has pitted resourceful scientists against an elusive quarry, whose traces are scattered across the electromagnetic spectrum. The first molecular fingerprints were spotted at visible wavelengths in 1936 by Walter Adams and Theodore Dunham *(below)*. But succeeding advances had to await the development of new instruments that could scan unseen ranges of light—ultraviolet, infrared, and microwave—for the spectral signs of the host of molecules lurking between the stars.

1936 At Mount Wilson Observatory, Adams *(below, right)* and Dunham began the hunt for interstellar molecules when they found unusual absorption lines, later linked to cyanogen and methylidyne, in stellar spectra.

1953 Charles Townes of Columbia University headed the team that first generated intense beams of microwaves with a maser *(bottom)*. The breakthrough helped astronomers in the 1960s to recognize cosmic maser emissions from molecular clouds.

earning a doctorate in physics from the California Institute of Technology in 1939, he spent the years of the Second World War at Bell Telephone Laboratories, designing radar bombing systems. Radar involves the generation and reception of microwaves, which fall in the millimeter to centimeter range of the electromagnetic spectrum. Townes's interest in these particular wavelengths led him to great advances in the emerging fields of quantum electronics and radio astronomy.

In 1953, Townes and the Columbia team succeeded in measuring the rest frequencies generated by the hydroxyl radical (OH), and two years later, Townes reported similar measurements for the molecules ammonia, carbon monoxide, and water. OH, for example, produced four closely spaced emission lines near the 18-centimeter wavelength—at frequencies of 1,612, 1,665, 1,667, and 1,720 megahertz (million cycles per second). As he noted at the time, the low temperatures that would cause molecules to generate low-energy radio photons corresponded to the presumed low temperatures of interstellar space. Townes's experiments revealed that each line was slightly different in intensity. The line at 1,667 megahertz, for example, was characteristically two times stronger than the line at 1,665.

With the theoretical groundwork in place, all that researchers hunting for interstellar molecules needed at this point were adequate tools for detection. That would take another decade, but by 1963, Sander Weinreb, Alan Barrett, and their group from the Massachusetts Institute of Technology had devised

1963 A group led by Alan Barrett *(below, left)* and Sander Weinreb of MIT used innovative techniques for the spectral analysis of microwaves to discover the absorption signature of interstellar hydroxyl.

1965 Harold Weaver and a team of Berkeley astronomers detected interstellar emission lines so powerful that Weaver dubbed the source "mysterium." It was later identified as a cloud of hydroxyl masers.

1970 George Carruthers of the Naval Research Laboratory designed the instruments for a rocket-borne probe that recorded the first traces of molecular hydrogen at ultraviolet wavelengths.

equipment and digital techniques that would enable them to conduct the necessary analysis of radio spectra at microwave frequencies. In that year they ended the drought of discovery.

During ten days in October, Weinreb's group observed Cassiopeia A, a powerful radio source in the constellation Cassiopeia. Using the eighty-four-foot parabolic antenna of MIT's Millstone Hill Observatory and Weinreb's new digital system of spectral analysis, the team found the absorption lines that matched the laboratory prediction for the hydroxyl radical. Judging from the lines' Doppler shift, the cloud of molecules lay somewhere between Cassiopeia A and Earth—a finding that set off a flurry of radio examinations of similar bright radio sources. And what awaited the scientists there was surprising indeed.

Two years after Weinreb's breakthrough, three groups of observers in the United States, looking at various radio sources in or near the galactic plane, detected microwave emissions at the OH frequencies of 1,665 and 1,667 megahertz. But the relative intensities of the lines were peculiar: The line at 1,665 megahertz, which was expected to show up two times weaker than the line at 1,667 megahertz, was three times stronger. Moreover, since observers had found no emissions at 1,612 and 1,720 megahertz, they could not be certain that the emitting source was in fact hydroxyl. A group at the University of California at Berkeley headed by astronomer Harold Weaver was initially so perplexed by the emissions that its members half-jokingly gave

1970 The roster of interstellar molecules expanded to include carbon monoxide (an index of large amounts of molecular hydrogen), thanks to keen detection at microwave frequencies by a trio of radio astronomers from Bell Telephone Labs: Keith Jefferts (below, right), Robert Wilson (center), and Arno Penzias, shown pointing to a source of carbon monoxide emissions on a transparent model of the sky.

1972 Berkeley's William J. Welch opened the radio telescope array at Hat Creek Observatory, whose multiple dishes sharpened resolution at the millimeter wavelengths characteristic of many interstellar molecules.

whatever was producing them the name "mysterium." As it worked out, the missing lines were detected within five years, but the unusual intensity ratios of the emissions remained baffling. It was as though the OH molecules had been excited by some outside source so as to amplify their radiation at a particular frequency.

FROM "MYSTERIUM" TO MASERS

By a bit of scientific serendipity, astronomers already had a terrestrial model for this phenomenon: It was a device known as a maser, an acronym for "microwave amplification by stimulated emission of radiation," and its inventor was Columbia's Charles Townes. (In 1964, Townes would share the Nobel prize for physics with a pair of Soviet scientists who independently came up with a similar device.)

As the story of the maser is told, Townes was sitting on a park bench in Washington, D.C., early one morning in 1951, pondering a vexing problem: how to build a device to generate very short-wavelength, high-intensity microwaves in order to obtain a sharper radar image. Mechanical oscillators that produced microwaves in the centimeter range were readily available, but scaling them down to emit microwaves of millimeter length presented seemingly insurmountable manufacturing difficulties.

Then Townes got the idea of using molecules, which radiate at specific frequencies, or wavelengths, whenever they drop from a higher energy state

1973 A team led by Patrick Thaddeus of Columbia University and the Goddard Institute for Space Studies mapped carbon monoxide emissions from an area in the Orion cloud complex thirty times larger than the face of the Moon.

1975 A theory that new stars emerge when molecular clouds are compressed and heated by shock waves was worked out at Harvard by Bruce Elmegreen *(below, left),* drawing on the observations of Charles Lada *(right).*

1985 Louis Allamandola of NASA Ames Research Center added a complex new class of organic molecules to the interstellar equation when he attributed emissions in the infrared range to polycyclic aromatic hydrocarbons.

to a lower one. In December 1953, Townes and his Columbia colleagues sent a beam of ammonia, with its molecules distributed between high-energy and low-energy, or ground, states, through an electrical field. Molecules in the high-energy state remained in the beam, and those in the ground state were drawn away. Subsequent exposure to another electrical field caused the swarm of high-energy molecules to drop to the ground state virtually simultaneously. The result of this stimulated emission was an intense beam of photons at one frequency. It was the first manufactured maser. And for all that Townes or anybody else knew, it was the only maser in the universe.

Within a few years, masers were being employed in a variety of devices, including radio telescopes, where they were used to amplify weak signals without introducing random noise. They were therefore familiar to the scientists studying interstellar OH emissions—who soon concluded that the extraordinary intensities of these radiations must be the product of nature's own version of Townes's invention. As more observations of OH emissions were gathered, researchers found them to be associated with regions of ionized hydrogen, or HII, as well as with young stars, which tend to emit primarily infrared radiation. Apparently, the hydroxyl molecules were being pumped up to higher energy levels by infrared radiation from nearby stars, allowing the OH to amplify the background radiation *(pages 96-103)*.

THE CRITICAL YEARS

Even as some researchers worked out the details of how interstellar molecules could act as cosmic masers, others continued to scan the radio universe for signs of more molecular species. In 1968, a team of observers from the University of California at Berkeley—including Charles Townes, as well as Albert Cheung, David Rank, Douglas Thornton, and William J. Welch—used the school's 20-foot radio telescope near Hat Creek in the northern Cascade Mountains to detect the microwave signature of the ammonia molecule at 1.25 centimeters in the direction of the constellation Sagittarius. Later that year, the same group found the emission lines of water molecules at 1.35 centimeters in three different regions of the sky: one in Sagittarius, toward the galactic center, that seemed to coincide with the previously discovered ammonia source; one toward the Orion nebula; and a third, designated W49, in the direction of the constellation Aquila. And in 1969, scientists working with the 140-foot radio dish at the National Radio Astronomy Observatory in Green Bank, West Virginia, brought the number of known interstellar molecules to six when they found formaldehyde in fifteen of the twenty-three sources they surveyed.

The discovery of formaldehyde, a fairly complex organic molecule, was particularly significant, for it opened up the possibility of finding the building blocks of terrestrial-type life in the vast regions between stars. Of the four basic gases that scientists theorize made up Earth's primordial atmosphere, three—hydrogen, water, and ammonia—radiate at radio wavelengths and have been detected in interstellar clouds. The fourth, methane, is not a radio emitter, however, and although it emits in the infrared in terrestrial laboratories, astronomers have not been able to detect its weak infrared signal in space. (Earth's atmosphere absorbs some wavelengths of infrared, but those at the far end of that portion of the spectrum are susceptible to detection by radio telescopes.) But formaldehyde is chemically related to methane, and its presence in interstellar space leads many observers to conclude that methane exists out there as well.

Scientific resistance to the idea that complex molecules could survive in interstellar space finally bowed to the growing accumulation of evidence. Clearly the density of gas and dust in the regions between stars was high enough not only to allow the formation of complicated molecular species but also to protect them from the destructive forces of ultraviolet radiation and cosmic rays. With the search joined in earnest, astronomers set a good pace in the new decade by detecting interstellar molecular hydrogen (H_2) and carbon monoxide (CO).

Both had been predicted for some time. As early as 1959, Lyman Spitzer and Franklin Zabriskie of Princeton suggested that interstellar H_2 might be detectable as absorption lines at ultraviolet wavelengths in stellar spectra. But since Earth's atmosphere blocks most UV radiation, that region of the spectrum poses a problem. In 1970, however, George Carruthers of the Naval Research Laboratory neatly sidestepped the obstacle by mounting his sensors—an electrographic spectrograph and ultraviolet photometers—on board an Aerobee-150 rocket. Launched from the White Sands Missile Range in New Mexico, the rocket rose high above the atmosphere, where it recorded narrow, sharp absorption lines in the ultraviolet spectrum of the star Xi Persei in the constellation Perseus. Carruthers's measurements indicated that nearly half of all the interstellar hydrogen in the line of sight to this star was in molecular form, a finding that agreed with theory.

Although Carruthers's achievement was a major breakthrough in detection, astronomers would have difficulty exploiting it. For one thing, rocket launches in the cause of astronomy are not everyday occurrences. For another, molecules of H_2 are symmetrical, meaning that each is made up of two identical hydrogen atoms. Molecular hydrogen thus does not undergo distinct energy state transitions that would produce a discernible radio spectral line *(pages 66-67)*.

A CONVENIENT ASSOCIATION

Scientists therefore looked for another way to trace the distribution of this molecule. The answer lay in a chemical companion: carbon monoxide (CO). Given that hydrogen is the most prevalent element in the universe, and carbon is commonly associated with hydrogen, astrochemists theorized that molecular hydrogen might coexist with certain energy levels of carbon monoxide in a ratio of about 10,000 molecules of H_2 to every one molecule of CO. If astronomers could locate CO in space, they could estimate how much hydrogen was in a given interstellar region. Happily, carbon monoxide is an asymmetrical molecule that radiates at different wavelengths, depending on the type of energy transition. Since it radiates in the infrared and the microwave, CO is somewhat easier to detect from Earth than molecular hydrogen.

In 1970, Robert Wilson, Arno Penzias, and Keith Jefferts, of Bell Telephone Laboratories, attached a sensitive new receiver adapted from digital communications technology to the National Radio Astronomy Observatory's (NRAO) thirty-six-foot dish located on 6,300-foot-high Kitt Peak, forty miles southwest of Tucson. Almost immediately, they identified carbon monoxide molecules at 2.6 millimeters, or 115 gigahertz, in the direction of Orion. Before long, teams at Bell Labs, NRAO, and Five College Observatory near Amherst, Massachusetts, among other observers, picked up the track of CO almost everywhere they pointed their instruments.

As scientists were at last beginning to realize, interstellar molecules are not just scattered randomly in space or even gathered into small clumps. Instead, they seem to exist in vast populations, formed into enormous clouds of matter spread all around the heavens.

One such cloud was confirmed in 1973 by radio astronomer Patrick Thaddeus of the Goddard Institute for Space Studies in New York, along with Kenneth Tucker and Marc Kutner of Columbia University. Working with the five-meter radio telescope at the McDonald Observatory, Thaddeus and his team spent fifty hours studying a patch of interstellar space near the Horsehead nebula in the constellation Orion. When they sampled 150 locations in the patch for CO, an astonishing picture emerged: The molecular cloud appeared to be 100 light-years in diameter, with 100,000 solar masses of gas and dust. As it turned out, the map could only hint at the real size of the cloud— paradoxically, because the McDonald telescope had too high a resolution.

The rule with both optical and radio telescopes is, the bigger the lens or dish, the higher the resolution but the narrower the field of view. The five-meter McDonald receiver could take in circular patches of the sky only 2.3 minutes of arc in diameter, or about one light-year across for an object at the distance of Orion. At that resolution, with a typical ten- or fifteen-minute exposure, the team would have needed years to map the full extent of the cloud. As Thaddeus later put it, "It was like trying to paint a barn with a quarter-inch brush."

To get a broader picture, Thaddeus had to build a bigger brush—which meant building a smaller dish. The team started work on a 1.2-meter instrument that could cover swatches of sky about eight arc minutes in diameter. The other advantage of the mini, as it was called, was that the scientists could dedicate it to the pursuit of carbon monoxide and not have to compete for time on the larger McDonald telescope. Two years later, Thaddeus installed the mini in what seemed a wildly inappropriate place: on top of the fifteen-story Pupin Physics Laboratories on the campus of Columbia University, in the heart of New York City. Like any modern metropolis, New York was awash in light and radio interference, and no professional astronomers had attempted optical or long-wavelength radio observations there in half a cen-

The map at right—showing the vast Orion cloud complex, with the dense areas of potential star formation coded red—was completed by a team of radio astronomers using a rooftop "mini" at Columbia University *(below, inset)*, a small 1.2-meter radio telescope with a wide field of view. Patrick Thaddeus and his associates began the task using the high-resolution, five-meter dish at the McDonald Observatory *(bottom)*. But the big dish's narrow field rendered a comprehensive survey of the region prohibitively time-consuming. After mapping a six-square-degree section of the Orion complex *(box at right)* with the McDonald dish, Thaddeus hit on the idea of the mini, which sped up the voluminous project considerably.

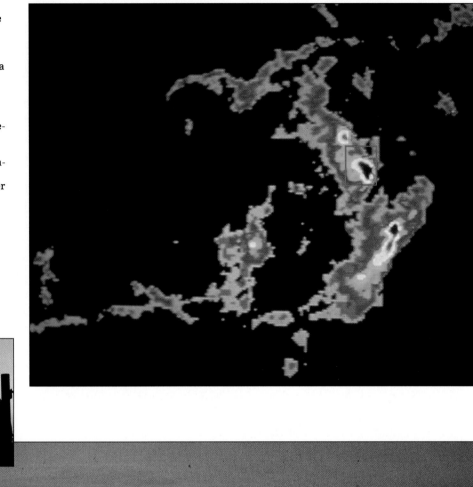

tury. But at the 2.6-millimeter band where carbon monoxide leaves its signature, the city was, in Thaddeus's words, "as quiet as the day Henry Hudson first sailed upriver."

Over the next five years, the astronomers trained the telescope on the Milky Way, adding to surveys conducted by other teams. As Thaddeus reported later, "The clouds were everywhere. Within two or three years of the discovery of CO it was clear that they were important new constituents of this galaxy, and probably every other spiral galaxy."

While the mapping project proceeded, an exact copy of the New York mini was installed in 1982 at Cerro Tololo, Chile, to chart the portion of the Milky Way visible only from the Southern Hemisphere. Four years later, Thaddeus moved to the Harvard-Smithsonian Center for Astrophysics in Cambridge, taking his team and his mini with him. Both instruments use a technique to produce a so-called superbeam. Instead of focusing on each spot for five- to ten-minute exposures, the computer-controlled telescope moves through a four-by-four array of points, devoting a minute or two to each. The computer then averages the data from each array, and by this procedure can cover a wide area of sky in a given time.

As of 1988, the astronomers and the twin minis had performed more than 31,000 individual observations covering more than 7,700 square degrees, or nearly one-fifth of the sky. With a larger device—such as the five-meter McDonald dish or the twelve-meter radio telescope at Kitt Peak—the project would have required many decades.

What the mapping shows is that the galaxy contains thousands of molecular clouds and that the majority of them are concentrated in the plane of the galaxy, forming a ring that lies between 12,000 and 25,000 light-years from the center. The stars and atomic hydrogen gas, along with the largest molecular clouds, form a vast spiral about 50,000 light-years out. Looking toward the center from Earth's vantage, or through the plane of the galaxy, the clouds seem to hug the center line, with the stars forming a kind of lattice or superstructure around them.

DELIVERY ROOMS FOR THE STARS

The discovery and mapping of molecular clouds in the 1970s and 1980s have raised a swarm of intriguing questions that astronomers are striving to answer. A primary question, of course, is how the individual molecules within these clouds are formed. One theory is that they are created by simple collisions between atoms in the gas. Another holds that they begin as atoms clinging to the surface of dust particles *(pages 68-69)*, which seems to be the process favored for the growth of more complex molecules.

Other questions revolve around the dynamics of cloud formation. In a hypothesis proposed by Frank Shu and his colleagues at the University of California at Berkeley, cloud formation is partly the result of instabilities in the interstellar magnetic field: As these instabilities cause the interstellar medium to become lumpy, the lumps tend to collect into clouds. Another theory proposes that gravity alone is enough to make the stuff between the stars coalesce; as these aggregations move through interstellar space, they collide with one another and gradually form ever larger clouds. The irregular shape of interstellar clouds leads some astronomers to believe they are fairly young, perhaps not more than 30 million years old, or only a fraction of the age of Earth; if they were much older, gravitational forces would have rendered them more spherical.

Despite many uncertainties, astronomers are sure of one thing: Molecular clouds are intricately involved in the birth of stars, a phenomenon of interest to Charles Lada of the University of Arizona in Tucson. Lada was a graduate student at Harvard in the early 1970s when the current picture of molecular clouds began to take shape, and the detection of carbon monoxide was cause for elation. From Lada's point of view, CO was one more tool for probing the clouds that seemed to be consistently associated with any evidence of star formation. For his thesis in 1975, Lada set out to map the carbon monoxide in a molecular cloud near the Omega nebula, a bright emission nebula in Sagittarius. What he found were the first tentative answers to questions of where and how stars are born.

Lada observed the Omega nebula using the five-meter dish at the University of Texas McDonald Observatory. Because he knew the Omega nebula contained newborn stars, he expected to find carbon monoxide distributed evenly in all directions from the nebula, a result that would fit the accepted view of emission nebulae as star-forming regions deep within molecular clouds. He saw something else entirely. At one side of the Omega nebula, a cluster of older O and B stars shone as pinpoints of light through the gauzy wisps of gas ionized by their intensive radiation. The carbon monoxide, however, was concentrated on the opposite side of the nebula from the bright O and B stars, stretching away into space. The nebula was not buried deep in the molecular cloud but, rather, was a blister on its surface. Lada also noticed infrared sources in the carbon monoxide concentrations, the marks of protostars, objects on their way to becoming stars.

As Lada described it, "The HII region, the infrared sources, and the cloud were all at the young end of the time arrow." He added, "I got this idea of a wave of pressure moving into the cloud with HII regions as the driving agents." That is, the radiation pressure of one generation of stars seemed to be spawning another, in a domino-like progression of star birth. The idea was not entirely new: In the 1940s, the Armenian astronomer Viktor Ambartsumian had found that some clusters of massive young stars resembled monumental bursts of fireworks. And in 1964, the Dutch astronomer Adriaan Blaauw had noted that certain groups of stars line up with the young ones at one end, the older stars at the other. Lada's study provided both a mechanism and a framework for these ideas.

In 1975, the same year that Lada presented his thesis, Bruce Elmegreen, a young astronomer from Princeton with the ink on his Ph.D. diploma not yet dry, settled in at Harvard as a junior fellow. "When Bruce heard what I had

CONTOURS OF A STARRY NURSERY

Deep within vast clouds of dust and gas, some of the universe's most dynamic events—from the birth of stars to the maelstrom of activity at the center of galaxies—take place unseen by even the most powerful optical telescopes. In the optical image at left, for example, obscuring patches of nebulous material in the vicinity of Rho Ophiuchi *(blue star, top)* give no indication of their role as a stellar breeding ground.

But with instruments tuned to other, less energetic forms of electromagnetic radiation, particularly infrared and radio waves, astronomers get a clear and penetrating view of these clouds and, by studying their structure, learn more about the processes at work inside them. The secret to the technique lies in the fact that interstellar clouds are typically composed of several varieties of complex molecules, each of which emits radiation under different conditions. By charting emissions from specific types of molecules, astrophysicists can create contour maps that trace patterns of heat, density, and turbulence within a cloud. In stellar wombs such as those of Rho Ophiuchi, these patterns hold clues to almost every developmental stage in star synthesis, from the initial condensation of diffuse material to the final collapse that ignites the nuclear fires and signals the moment of star birth.

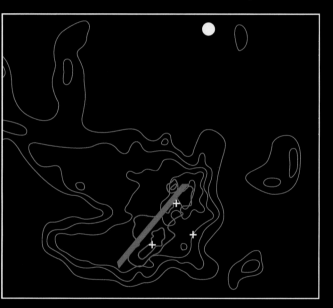

Patterns of density. A radio image of the boxed area on the opposite page shows the distribution of two different isotopes of carbon monoxide in the cloud near Rho Ophiuchi *(yellow dot)*, with contour lines denoting increasing density. One of the isotopes *(blue)* emits radiation from throughout the cloud and thus gives an idea of its overall shape; the other *(purple)*, which is detectable only at higher concentrations, maps a dense interior region. The areas of greatest density suggest the presence of a compression front *(red line)*, where a shock wave—perhaps from a supernova—has pushed interstellar material closer together. Infrared sources *(yellow crosses)* are new stars that formed after the front passed through.

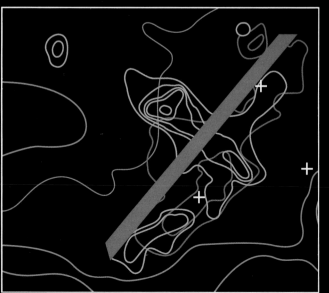

A potential site for new stars. An enlarged view of the area near the compression front includes readings from another molecule *(orange contour lines)*, which emits discernible radiowave energy only when the molecules are concentrated in cold, dense clumps. Unlike the patch of carbon monoxide *(purple)* that was condensed by the pressure of the shock wave moving in from the lower right, portions of this compound are already quite dense well ahead of the compression front. Astronomers predict that as the shock wave storms through, it may further compress these thick cores of gas, precipitating the formation of new stars.

Probing the Heart of the Galaxy

The Milky Way teems with molecular clouds, as is evident from the map below, which charts radio emissions from carbon monoxide throughout the galactic plane. The increasing intensity of the signals, represented by the range of colors from blue to green, yellow, red, and white, indicates that these clouds accumulate toward the galaxy's core *(arrow)*. Some astronomers speculate that at the very heart of this murky region lies a prodigious black hole, its mass equivalent to millions of Suns and its gravitational force so powerful that nothing within several million miles, not even light, can escape it. The only way to confirm such an object's presence is by its effects on matter surrounding it; the clustering veil of molecular clouds may thus hold the secret to the mystery.

The best information about the innermost clouds comes from molecules of hydrogen cyanide (HCN), which provide especially detailed pictures of the clouds' structure and dynamics. As illustrated by the two maps at bottom, HCN emissions reveal distribution patterns and orbital velocities consistent with the theory of a central black hole.

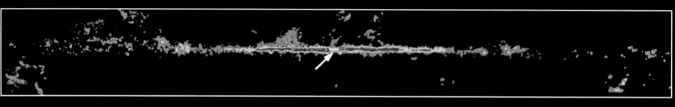

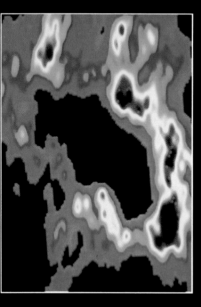

Girdling clouds. This color-coded image *(left)* of hydrogen cyanide emissions shows an elliptical ring of molecular clouds fifteen light-years in diameter surrounding the center of the galaxy *(arrow, above)*. Lavender represents the thinnest concentrations, dark red the densest. The irregular clumpiness suggests great turbulence, perhaps caused by the violent demise of a star that thousands of years ago wandered too near a central black hole.

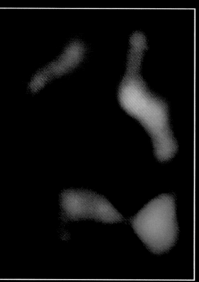

Molecules in motion. Here, color coding denotes the molecular ring's motion away from Earth *(red, yellow)* and toward it *(blue, green)*. Measurements indicate an orbital velocity of more than 100 miles per second. The rate at which this speed decreases with distance from the galactic center has enabled astrophysicists to estimate the center's mass at a few million Suns.

Dissecting a Starburst Region

At a wide range of electromagnetic wavelengths, the galaxy known as M82, 10.6 million light-years distant, presents a picture of intense activity. Optical images such as the one below reveal great disturbances and irregular patterns throughout the galaxy's hazy disk of stars. Infrared observations delineate a small region near its center where numerous massive stars once formed, eventually exploding as supernovae. Radio emissions from a ring of molecular gas near this central region confirm the strange dynamics and have helped lead scientists to some tentative conclusions.

The pressure in these gaseous clouds, determined from temperature and density calculations derived from two types of molecules, is more than ten times greater than it is in a similar central region of the Milky Way and may be triggering further massive star formation within the molecular ring. Astrophysicists speculate that such high pressure is the result of gravitational interactions with a neighboring galaxy, M81. The interactions could have caused gas in M82's outer regions to fall toward the center, initiating the so-called starburst episode, which in turn forced any remaining gas into the compressed ring visible on radio maps.

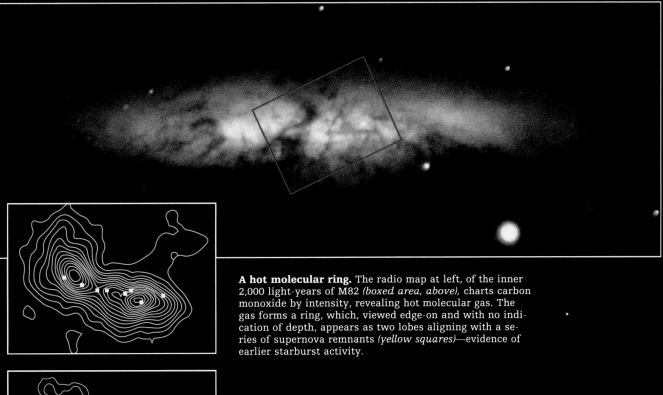

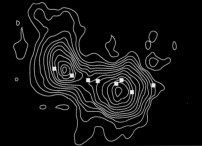

A hot molecular ring. The radio map at left, of the inner 2,000 light-years of M82 *(boxed area, above)*, charts carbon monoxide by intensity, revealing hot molecular gas. The gas forms a ring, which, viewed edge-on and with no indication of depth, appears as two lobes aligning with a series of supernova remnants *(yellow squares)*—evidence of earlier starburst activity.

The density component. Readings from formyl ions, which radiate only under conditions of extreme density, exhibit the same ring structure as for carbon monoxide, again aligned with the supernova remnants. The initial starburst activity probably occurred in the ring's center, compressing and heating the encircling band of gas and creating pressures high enough to spark additional starburst episodes moving continually farther out.

found, he got very excited," recalled Lada. "Soon he had the whole theory worked out." Elmegreen wove an intricate theoretical model that explained what Lada had seen in the Omega nebula. As Elmegreen figured it, "The intense radiations produced by the first generation of O and B stars are heating up the surface of the molecular cloud to temperatures around ten thousand degrees, which drives a shock wave into its depths. This shock scoops up the gas like a snowplow, creating dense clumps that soon contract and ignite into the next generation of stars."

The Lada-Elmegreen model, introduced at an international conference in Geneva in 1976, sent its own shock wave of excitement through the astronomical community. The two young scientists had arrived at a model for star formation that is still in use today, although it does not explain every aspect of the process. Astronomers believe, for example, that at least some fraction of average stars such as the Sun are born in the interiors of molecular clouds, rather quietly, away from the cataclysms taking place at the edges. Furthermore, magnetic fields and stellar winds—the currents of charged particles streaming outward from stars—are thought to play a part in the miracle of star birth, as do supernovae. Stars that end their lives as supernovae are so massive that they can burn up their fuel in less than three million years. As a result, they never get far from the cloud that gave birth to them, and when they explode they return a shock wave that can compress the cloud's gas and dust into new star-forming regions.

Despite the apparent straightforwardness of this scenario, Harvard's Thaddeus cautions, "The problems of star formation are wonderfully complex. In a place like Orion you can see the whole region disintegrating. Supernovas, stellar winds, shocks are slamming into the molecular cloud, sweeping it back, igniting a conflagration. Things are happening wholesale. You have to be careful in applying simple models."

A SHARPER VIEW

Today, astronomers are assiduously probing the depths of molecular clouds in search of new and exotic molecules. A state-of-the-art instrument for such work is the Berkeley-Illinois-Maryland Array at Hat Creek, operated by the University of California at Berkeley. The creation of William J. Welch, director of Berkeley's Radio Astronomy Laboratory, the telescope consists of three twenty-foot antennas, with six more scheduled for completion by 1992. Linked together, the nine instruments will simulate a single dish 1,000 feet in diameter. Known as an interferometer because it takes advantage of the way radio waves interfere with each other, the array will have a resolution, at millimeter wavelengths, of one one-hundredth of a light-year at the distance of Orion—roughly equivalent to distinguishing a crater two miles wide on the Moon.

In 1987, a team of Berkeley astronomers led by Reinhard Genzel, now director of the Max Planck Institute in Bonn, trained the Hat Creek interferometer on the galactic center, in the direction of Sagittarius. They were searching for hydrogen cyanide (HCN), a simple straight-line configuration of hydrogen, carbon, and nitrogen that is common in laboratories on Earth. Hydrogen cyanide had been detected in relatively high abundances near the core of the galaxy, and the team was looking to plot its location and behavior in an effort to study the dynamics at the galactic core. The group found the molecule concentrated in a turbulent ring around the galactic center, highly inclined to the galactic plane. Another team, using the Very Large Array in Socorro, New Mexico, observed what are thought to be streamers of ionized hydrogen cyanide falling inward from the edge of the ring. This team also determined that the gas circles the galactic center at about 100 miles per second. These speeds, as well as the high temperature of the gas—400 degrees Kelvin, compared with the 10 to 15 degrees characteristic of gas farther from the center—add substantially to evidence for the existence of a massive black hole at the heart of the Milky Way.

So-called ring molecules are also prized quarry in the interstellar hunt, if only because their complexity holds out hope for finding the building blocks of terrestrial life in extraterrestrial locations. Carbon rings like benzene (C_6H_6), for example, are of vital importance in organic chemistry, and although no traces of benzene have yet been found, scientists have found other ring molecules. Thaddeus and his team detected silicon dicarbide (SiC_2) in 1984 and cyclopropenylidene (C_3H_2) in 1985, the first findings of relatively stable, complex interstellar molecules rather than simple chains.

One of the most vigorously pursued classes of ring molecules is known as the polycyclic aromatic hydrocarbons, or PAHs, which can consist of a dozen or more carbon rings or hexagons *(pages 72-73)*. Certain PAHs, such as the poisonous ones that are found in auto emissions and the soot from coal-burning fires, are all too plentiful on Earth. None have yet been positively identified in space, but experimental astrochemist Louis J. Allamandola of the NASA Ames Research Center at Moffett Field in California reported in 1985 that he detected traces of what he believes are PAHs in the interstellar medium. Some astronomers estimate that these compounds could account for as much as 10 percent of all the carbon in the galaxy. Perhaps one of the more intriguing aspects of PAHs is that they are an important link in the chain of formation between interstellar gas—which is to say, atoms and molecules—and interstellar dust.

BURGEONING COMPLEXITY

As more and more complex organic forms are added to the roster of interstellar molecules, speculation about the role of these molecules in the evolution of life on Earth increases accordingly. By far the most extreme theory has come from astronomers Sir Fred Hoyle and Chandra Wickramasinghe, who proposed in 1979 that infectious diseases are the result of invasions of interstellar viruses and bacteria, that human beings are the offspring of living cells brought to Earth by comets, and that insects are, in fact, creatures from outer space.

This theory is largely dismissed by the scientific community, whose members would be delighted, at this point, to find evidence of amino acids in interstellar space. These molecular chains are the building blocks of proteins, which in turn are the precursors of biological life. Amino acids have been found in meteorite fragments, leading some researchers to argue that they are of extraterrestrial origin. But other, more cautious or skeptical scientists suggest that the presence of acids is only due to earthly contamination. So far, no amino acids have been discovered in molecular clouds, and even if they were to be—an event that many astronomers think is likely —the finding would not necessarily confirm a connection between organic matter in molecular clouds and that found on the planet's surface. Most scientists agree that if any amino acids were developing within the cocoon of giant molecular clouds, they would be destroyed as soon as they emerged from the protective folds.

Still, the search for ever more complex and intriguing interstellar molecules goes on, and many astronomers feel the possibilities are endless. As Harvard's Patrick Thaddeus has put it, "My guess is that almost everything is there at some level. I'm prepared to believe almost any reasonable molecule can be made in space." Boston University astronomer Thomas M. Bania agrees, pointing out that the same laws of physics apply, whether the organic chemistry in question is earthbound or interstellar. "If you give the universe half a chance," Bania has said, "it drives itself toward complex, molecular, organic systems. And we know from one experiment, namely us, that if these organic systems get complex enough, they start writing poems and building cities."

MOLECULAR AMPLIFIERS

Although a fledgling star can be ten times hotter than the Sun and a dozen times as large, it may remain hidden from optical view by the dense cloud of gas and dust from which it has emerged. Fortunately, there is another way that such sheltered objects can announce their presence. The placental cloud that envelops an embryonic star is an ideal environment for the formation of molecules—particles consisting of two or more atoms. As these molecules accumulate, they tend to become excited; they may then release their built-up energy in a beam of microwave radiation that alerts radio astronomers to the new star's location.

The wavelengths detected by radio telescopes identify the molecules that generate the signals—often such ordinary compounds as water (two hydrogen atoms bound to one oxygen atom) and hydroxyl (one hydrogen atom bound to one oxygen atom). Because the molecular swarms are huge, their microwave emissions are extraordinarily intense, having been amplified billions of times by the excited particles. Such generators of strong microwave radiation are known as masers, an acronym for "microwave amplification by stimulated emission of radiation." A maser is analogous to a laser (light amplification by stimulated emission of radiation) except that its concentrated beams are invisible. Cosmic masers are far more powerful than those produced in laboratories. A water maser in the constellation Aquila, for example, emits energy roughly equivalent to the total electromagnetic output of the Sun.

Once viewed as a cosmic curiosity, masers are now considered vital astronomical tools. Besides revealing regions of star birth, they can be used to gauge the distance to remote points in space or to chart the convoluted magnetic field of the Milky Way galaxy *(pages 102-103)*.

How Masers Turn On

The remarkable capacity of interstellar clouds to dispatch detectable signals to Earth is explained by the laws of quantum physics, which state that molecules, like the atoms they are made up of, can exist only in discrete energy states. Normally, a molecule remains at the lowest energy level, called the ground state. Under certain conditions, however, a molecule may be pumped to a higher state, causing it to become excited—that is, to spin or vibrate faster. Astronomers speculate that there are two such pumping mechanisms at work in molecular clouds: collisions between molecules, which transfer energy as illustrated at lower right, and incoming radiation from nearby stars. In either case, the excited molecule gains a specific quantum of electromagnetic energy. When the molecule reverts to a lower state, or de-excites, it surrenders that energy in the form of a photon, a unit of electromagnetic energy whose wavelength is the same for all identical molecules undergoing the same reversion. In an interstellar cloud, billions of water molecules, for example, may de-excite in conjunction and emit a stream of photons at the same wavelength and in the same direction to yield the highly amplified signals picked up by radio telescopes.

This process is unfolding to dramatic effect in a star-forming region within the Orion nebula (outlined in black on the photograph at right). This area appears bright because radiation from more mature stars has gradually dissipated the surrounding cloud of molecules and also ionized its atoms, thereby producing a glow. But lurking amid the bright patches are dark areas that are dense with masers, allowing radio astronomers to pinpoint fledgling stars that are cloaked from view by the molecular shroud.

The evolution of a cosmic maser begins when molecules become excited. In this case, the excitement is generated when a diatomic molecule of hydrogen—the main constituent of interstellar clouds—collides with a water molecule. (The hydrogen atoms here are blue; the one oxygen atom in the water molecule is green.)

As a result of the collision, energy is transferred to the water molecule, which jumps to a higher energy state and spins faster. Conversely, the hydrogen molecule loses energy.

Soon, the excited molecule spontaneously drops to a lower energy level, emitting a photon *(purple)*. A few molecules drop to the ground state, but most, such as the one shown here, descend to an intermediate level called the metastable state, where vast numbers of them can remain for a relatively long time.

When a passing photon *(below, right)* strips a metastable molecule of its energy, it drops to the ground state and emits a photon of the same wavelength *(below, left)*, which travels in the same direction as the impinging photon. The two photons encounter other metastable molecules, reinforcing the maser.

A Hotbed of Molecular Emissions

The most powerful masers in the Orion nebula are produced near a phenomenally energetic object known as IRc2 (short for "infrared compact source number 2"). As its name implies, IRc2 announces its presence on far-infrared images, although at optical wavelengths its light is effectively shrouded by the surrounding cloud of dust and gas. Indeed, radio and infrared observations indicate that IRc2 is a massive young star emitting as much as 100,000 times the energy of the Sun and losing mass a billion times more rapidly. An infrared photograph of the same star-forming region outlined on page 96 shows a ruddy butterfly-shaped nebula *(inset)*, whose two wings are thought to be lobes of dust warmed by IRc2.

By analyzing masers generated in this area, radio astronomers have drawn up a diagram of molecular activity around IRc2, which is superimposed on the infrared photograph and detailed opposite. Circling the star is a torus of dust interspersed with hydroxyl and water molecules. The Doppler effect—the change in frequency of electromagnetic waves as the source emitting them moves toward or away from the receiver—has allowed astronomers to determine that the torus is revolving in a clockwise direction. Maser emissions spawned on the right-hand side of the ring are compressed to a shorter wavelength, and are thus detected at a higher frequency, because the molecules producing them are approaching Earth (a phenomenon known as blue-shifting). By contrast, maser emissions on the left-hand side are longer in wavelength—and lower in frequency—because the particles are receding (red-shifting). In addition to this rotating ring of molecular signals, water masers emit radiation from the center of the torus at either pole.

The masers around IRc2 can be extraordinarily intense. In 1985, for example, one water maser in the torus became so dominant that it interfered with other radio telescope observations at that wavelength within several degrees of IRc2.

Radio observations of IRc2 show that masers are being generated throughout a doughnut-shaped ring girdling the new star. The inner part of this gradually expanding torus contains mostly hydroxyl molecules; the outer part contains some water molecules as well. By tracking the Doppler shifts of the maser signals, astronomers have found that the torus is rotating clockwise. (The area moving toward Earth is shaded blue, and the receding section is shaded red.) Shooting out along the rotational axis of the torus are high-velocity water masers.

99

A New Yardstick to the Stars

Astronomers are constantly seeking more accurate ways to measure distances to stars and galaxies. Traditionally, celestial range finding often involved gauging stellar luminosity, a calculation that could be thrown off by the dimming effect of interstellar dust. Recently, astronomers have devised ways to avoid such optical snags by using masers as yardsticks.

One complex method of determining distance with masers, detailed below, rests on the simple premise that molecules functioning as masers exhibit certain random motions that can be measured and compared. To estimate the distance of a molecular cloud from Earth, astronomers take a sample of masers within the cloud and graph the distribution of two such variables: radial velocity (the speed, as revealed by the Doppler shift, at which the maser source is traveling toward or away from Earth) and lateral velocity (the movement of the source across the sky over time, as measured in fractions of a degree per year). Theoret-

Measurements Based on a Narrowing Curve

In theory, if a radio telescope were operating within a molecular cloud (above, left), it would detect the same distribution for the lateral velocities of the masers (purple) as for their radial velocities (red); when graphed, the two variables would yield identical bell-shaped curves. In practice, however, the observed distribution of lateral velocities narrows over distance; the farther away the molecular

cloud, the tighter the arc traversed by the radio telescope (yellow beam). Since distance has no such effect on the distribution of radial velocities, astronomers need only translate the difference in width between the two curves (above, right) into a ratio to calculate how far away the molecular cloud is. The entire process requires extraordinarily fine measurements, achieved by electronically

linking radio telescopes at widely distributed locations to heighten the resolution of observations. To gauge the distance to one nearby spiral galaxy, for example, astronomers must measure the lateral motions of masers that are crawling across the sky at an annual rate of just five-millionths of an arc second—or about the width of a human hair viewed from a distance of 2,500 miles.

ically, the graphs should always look the same, since the movements of the molecules are random and should even out radially and laterally. But in fact, the lateral distribution narrows markedly as the distance of the molecular cloud from Earth increases. This narrowing is a trick of perspective produced by the fact that a telescope sweeping a given field traverses a smaller arc when the field is farther away. Yet the trick is revealing, because by comparing the observed distribution of lateral velocities to that of radial velocities—a factor that is unaffected by distance—scientists can determine how far the region is from Earth.

Astronomers succeeded in confirming the validity of this procedure by tracking clumps of masers located in the Orion nebula; the distance estimate of 1,600 light-years agreed closely with the results of other methods. A related technique was then applied to maser emissions near the center of the Milky Way (outlined in black below). The distance arrived at—approximately 23,000 light-years—is an important benchmark, because it can be used to determine the galaxy's mass and gauge distances to other galaxies, leading to more accurate estimates of the size and age of the universe.

Mapping a Maser's Magnetic Field

Until recently, interstellar magnetic fields were largely ignored by astronomers. Such remote fields were notoriously hard to chart, and their significance was a matter of conjecture. Astronomers are now convinced, however, that nearly all cosmic matter and radiation are influenced by magnetism. Detailed knowledge of magnetic fields is thus essential to interpreting the radiant information from distant stars and nebulae. Fortunately, these once-hidden fields are yielding their secrets, due in part to an elegant new technique that uses cosmic masers as compasses.

Central to this technique is a phenomenon known as the Zeeman effect: the capacity of a strong magnetic field to split the energy levels of a molecule, causing it to emit electromagnetic signals not at one characteristic frequency but at a few slightly different frequencies. A maser passing through such a field will be

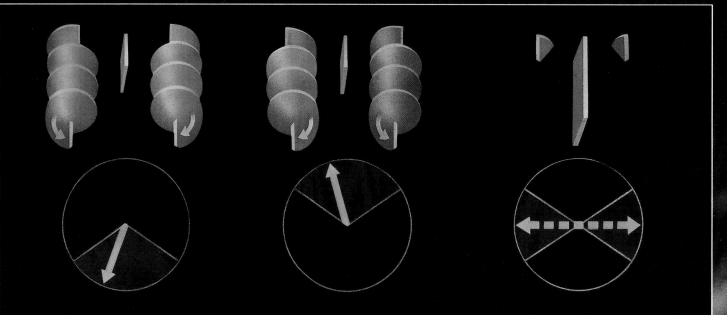

To map a magnetic field, astronomers compare the frequencies of the two outer signals. Here, the right circularly polarized signal *(clockwise spiral)* has a higher frequency than the left circularly polarized signal; both are strong relative to the linearly polarized signal *(plane)*. Drawing upon the demonstrated behavior of molecules emitting light waves in a magnetic field, researchers know that the field's north magnetic pole points toward Earth, somewhere in the light gray zone.

In this example, the maser's right circularly polarized *(clockwise)* signal has a lower frequency than the left circularly polarized signal. (The two spirals are transposed here to mark the shift.) This means that the magnetic field is reversed and that the field's north magnetic pole points away from Earth. The stronger the circularly polarized signals are relative to the linearly polarized signal, the closer the north magnetic pole comes to pointing directly along the line of sight.

When the linearly polarized signal is stronger than the two circularly polarized signals, the poles of the magnetic field lie roughly perpendicular to the line of sight from Earth—somewhere in the two light gray zones. In such cases, the circularly polarized signals are so weak that astronomers cannot determine their direction, making it impossible to figure out which magnetic pole is north.

picked up by a radio telescope as three or more signals centered where the lone signal would otherwise be expected. Strong hydroxyl masers provide the most revealing example of this phenomenon, for they break up into three signals that lend themselves neatly to analysis *(box)*. The two outer signals, having been contorted by the magnetic field, are circularly polarized—that is, they travel in spirals, one curving to the right and the other to the left. (The direction of the spin is determined by running the signal through an elaborate filter that allows only those waves polarized in a given manner to pass through.) The third, intermediate signal is linearly polarized; it oscillates in a plane. The distinct character of the three signals enables a researcher to determine their arrangement and relative intensities, factors that reveal the magnitude and orientation of the magnetic field polarizing the maser.

As radio astronomers compile such compass readings in star-forming regions across the swirling galaxy *(below)*, they are charting the way for a new understanding of the relationship between magnetic fields and the rise of distant suns.

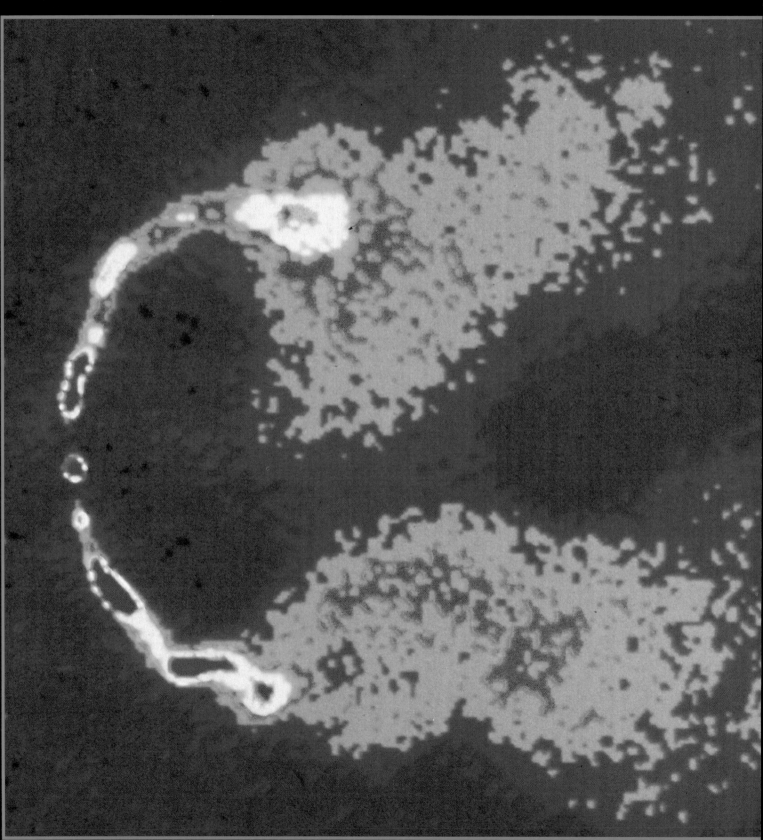

The swept-back plumes of radio emission from NGC 1265 offer astronomers indirect evidence of the intergalactic medium. As this so-called head-tail galaxy travels around the center of the Perseus cluster at a rate of 1,200 miles per second, its outpouring of radio energy meets unseen resistance—an ocean of diffuse gas that causes the radio signals to trail behind the galaxy like windblown hair.

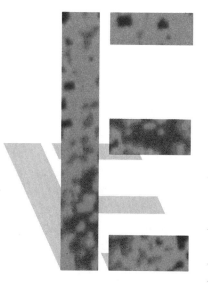ven as astronomers in the mid-1930s were confirming the existence of an interstellar medium, Leiden Observatory's Jan Oort stumbled on a mystery that would prove to be of intergalactic significance. In 1932, Oort was following up earlier work on the rotational dynamics of stars in the Milky Way with further observations designed to assess the amount of mass in a volume of space around the Sun. His method relied on the fundamental laws of gravity and motion. Stars orbiting the Milky Way's center tend to stray from the midplane of the galactic disk until the gravitational tug of all the stars above or below them pulls them back. The farther from the plane a star has wandered, the faster it travels as it returns toward the midline and the more it overshoots the mark. As a result, the stars describe a slow oscillation through the disk. Oort analyzed the displacement of absorption lines in the stars' spectra to estimate the speeds at which the stars were moving and calculated the gravitational force acting on them. Once he knew the force, he could determine the mass needed to account for it.

The answer was something of a surprise: Oort found a large discrepancy between the mass he had just calculated for the region and the sum of the masses of individual stars estimated from a formula known as the mass-luminosity ratio. The M-L ratio has been worked out by plotting the mass and luminosity values for a large number of binary star systems. First, the mass of each star in the pair is determined by observing its gravitational pull on its companion. Then each star's light is measured and related to its mass. Making this measurement for many binaries produces a graph that can be used to derive an estimate of the mass of a single star when only its luminosity is known. Astronomers use the Sun—an average star—as a standard, setting the ratio of its mass to the light it gives off at 1. Dim stars, such as white dwarfs, which are about the same mass as the Sun, can have M-L ratios of several hundred. Oort's motion studies suggested that there was twice as much mass in the galaxy than the mass derived from the M-L ratio. In other words, a substantial amount of the matter needed to explain the stars' motions was invisible at optical wavelengths.

Although the finding was puzzling, it was not entirely new. About a decade earlier, England's Sir James Jeans and Oort's mentor Jacobus Kapteyn had found comparable discrepancies between mass as determined from velocity studies and that derived from M-L ratios. But Oort and most of his contemporaries felt little concern, assuming that the unaccounted-for mass resided in small, dim stars, whose light simply was not detected with available telescopes, and in diffuse clouds of interstellar gas that were discernible only as stationary absorption lines in stellar spectra. As soon as scientists discovered more about dim stars and interstellar gas, Oort believed, the discrepancy would disappear.

A year later, however, Fritz Zwicky of the California Institute of Technology conducted similar studies from Mount Wilson Observatory and found even larger anomalies. Instead of nearby stars, he examined a vast group of galaxies known as the Coma cluster. Located about 400 million light-years away, the cluster contains thousands of galaxies gravitationally bound to each other and orbiting a common center of mass.

As Oort had done with stars, Zwicky estimated the range in orbital velocities of about 800 galaxies and calculated the mass needed to jibe with their motions and to keep them from flying apart. Then he summed up the individual galaxies' visible light to derive a mass estimate using the standard M-L ratio. The difference was substantial: The mass derived from the galaxies' velocities was 10 to 100 times more than the mass calculated from their light.

Unlike Oort, who remained unperturbed by these mass discrepancies, Zwicky made much of his results, declaring that, at least in the Coma cluster, "Much more dark matter exists than luminous matter." Since this notion directly contradicted the prevailing view that stars held the bulk of the material in the universe, many astronomers simply attributed Zwicky's finding to measurement errors that might have skewed the mass estimates.

A contour map of radio emissions from NGC 1265 *(pages 104-105)* reveals how the galaxy *(at red arrow)* is dwarfed by twin jets of high-speed gas more than 300,000 light-years in length. The jets have been deflected backward by ram pressure from extremely hot gas permeating the regions between galaxies in the Perseus cluster.

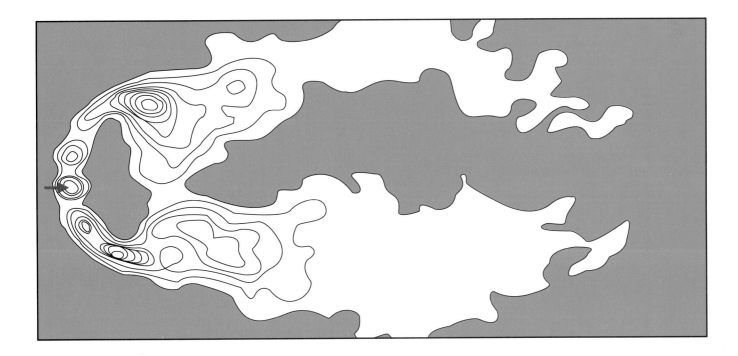

That Oort's observations suggesting the existence of invisible matter in the Milky Way were accepted as real—albeit not particularly noteworthy—while Zwicky's conclusion regarding galaxy clusters was dismissed out of hand may say more about the personalities of the two men than about the relative merits of their work. Zwicky, who was a professor at Caltech from 1925 until his death in 1974, was sometimes caustic and frequently controversial. His habit of spinning off radical theories—whether based on his own work or that of others—alienated him from many of his colleagues, and his temperament often cost him access to the big telescopes on Mount Wilson and Palomar Mountain. Consequently, he did most of his work on an eighteen-inch tele-

THE INTERGALACTIC SEARCH

The idea that matter might exist outside of galaxies began with an anomalous finding: There was not enough mass in all the stars visible in the Milky Way to account for their observed motions. When similar studies were carried out for other galaxies, and then for clusters of galaxies, the discrepancy only grew worse. As their instruments improved, scientists began to find signs of discrete gas clouds between galaxies, as well as signs of diffuse gas within clusters—but the missing mass problem, as it is known, remains unsolved.

1932 Jan Hendrik Oort studied the motions of stars through the plane of the Milky Way and found that the stars' mass as determined from their luminosity accounted for only 50 percent of the mass needed to explain their velocities.

1933 Fritz Zwicky's lifelong search for intergalactic matter began when he discovered that the discrepancy between mass and luminosity in the Coma cluster was even greater than that in the Milky Way.

1965 James Gunn (below, left) and Bruce Peterson reasoned that the lack of a broad trough of so-called Lyman-alpha lines in quasar spectra reflected the absence of a diffuse intergalactic medium of neutral hydrogen.

scope. Despite these handicaps, he made significant contributions in several fields, including the study of neutron stars, supernovae (he coined the term), and the large-scale distribution of galaxies, helping to compile a catalog of some 10,000 galaxy clusters in the Northern Hemisphere.

One colleague who did take Zwicky's pronouncements on intracluster matter seriously was physicist Sinclair Smith, a Chicago-born widgeteer with a Caltech doctorate who made various types of instruments for Mount Wilson's physics lab as well as for the observatory's telescopes. In 1936, Smith found a mass discrepancy in the Virgo cluster comparable to the one Zwicky had found with Coma. As it happened, however, Smith died two years later of cancer, and no one else took up the question. At the time, astronomers simply assumed that better telescopes would eventually discern what was by now known as dark matter—so called because it was detected only through its gravitational effects rather than electromagnetically.

Despite this lack of excitement, the issue of dark matter was becoming inextricably bound up with emerging ideas about the origin and fate of the universe. In 1929, Edwin P. Hubble, an American astronomer at Mount Wilson, had proved that the fuzzy "nebulae" once regarded as part of the Milky Way were actually galaxies in their own right, located very far away from the Sun's home galaxy. Moreover, these other star systems were receding from the Milky Way at speeds directly proportional to their distance. Hubble's discovery supported the idea that the universe was uniformly expanding in all

1966 Margaret Burbidge found heavy-element absorption lines in a quasar spectrum and theorized that the absorbing gas was associated with the quasar rather than located in intergalactic space.

1969 John Bahcall (below, left) and Lyman Spitzer proposed that gaseous halos around intervening galaxies accounted for some of the heavy-element absorption lines in quasar spectra.

1969 Vera Rubin and Kent Ford, finding that stars at the edges of spiral galaxies rotated faster than predicted by the laws of motion, theorized that the galaxies were embedded in huge halos of unseen, or dark, matter.

directions as a consequence of its explosive birth—known as the Big Bang—about 18 billion years ago.

But an expanding universe theory (in contrast to a "steady state" theory, which supposes the cosmos is static) raises the fundamental question of how the universe will end. On one side is the possibility that the universe will simply continue to expand forever. The opposite scenario is that it will eventually stop expanding and begin to collapse back on itself. Which outcome will prevail hinges on how much matter the cosmos contains. In the 1930s, astronomers calculated the so-called critical density—the amount needed to close the universe—to be approximately 2×10^{-29} grams in each cubic centimeter of space, or about ten hydrogen atoms per cubic meter. And by the end of that decade, it was clear that luminous matter, namely stars, accounted for only about 2×10^{-31} grams per cubic centimeter, or approximately one percent of critical density. Cosmologists regard this as a "missing mass" problem because Einstein's general theory of relativity predicts that the universal expansion will slow down due to gravitational forces if the density of matter in the universe is greater than critical density.

GRAPPLING GALAXIES
In the mid-1930s, scientists were only beginning to grasp the magnitude of the problem, but Zwicky stuck to his claims. Although he was occupied with research on many fronts, he continued to explore the dark matter question.

1971 Roger Lynds attributed the Lyman-alpha forest—densely packed hydrogen absorption lines in quasar spectra—to sparse clouds of gas, the first strong evidence for intergalactic matter.

1971 Herbert Gursky discovered x-ray emissions from clusters of galaxies in data from the *Uhuru* satellite and concluded that the radiation came from extremely hot, ionized gas between the galaxies in the clusters.

1973 James Peebles (*below, left*) and Jeremiah Ostriker used computer models to confirm Rubin and Ford's calculations that spiral galaxies must be embedded in a halo of dark matter in order to remain stable.

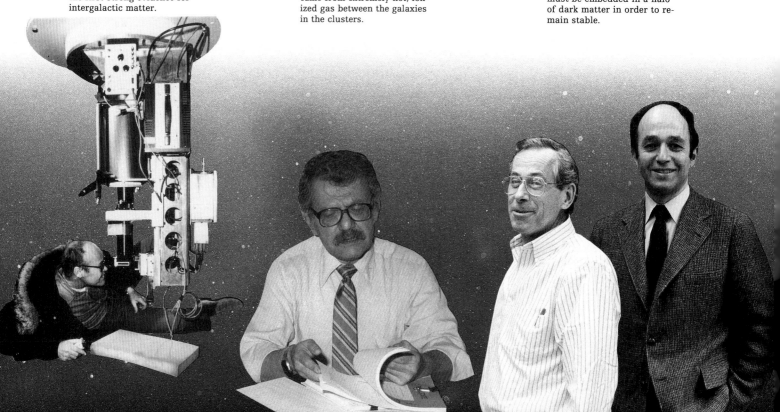

In a paper published in 1937, he suggested that galaxies grouped in clusters are so close to each other that they probably interact from time to time. These cosmic collisions and near misses would generate sufficient gravitational force to pull dust and gas out of the star systems and into the intervening space. Enough material might be lost to the regions between galaxies to make up the mass needed to stabilize their motions.

About fifteen years later, in 1952, Zwicky reported the results of a concentrated study of galaxies in the Coma cluster carried out with a new 48-inch telescope on Palomar Mountain. In the course of examining his own photographic plates and others made with the powerful 200-inch Hale reflector at Palomar Observatory, he had detected faint streams of stars, tens of thousands of light-years long, that seemed to run between galaxies. He characterized the streams as remnants of galactic close encounters, a finding that jolted some of his star-watching colleagues into mounting their own searches for galactic interactions.

As others took up the quest during the 1950s and 1960s, they began to catalog hundreds of so-called peculiar galaxies, odd-shaped stellar systems that possessed appendages variously known as "antennae," "tails," "plumes," and "bridges." Further support for Zwicky's insight came from computer models devised in the 1970s: Many peculiar galaxies, complete with stellar tails and bridges, can be re-created in computer simulations of galactic interactions. A decade or so later, even the Milky Way was found

1976 Richard Mitchell *(below, left)* and Peter Serlemitsos independently detected highly ionized iron in galaxy clusters, indicating that intracluster gas is not primordial but a product of stellar evolution.

1980 Wallace Sargent classified three sources of absorption lines in quasar spectra: intervening Lyman-alpha gas clouds, gaseous halos around intervening galaxies, and gas spewed from quasars.

1983 Stephen Schneider stumbled across a vast, dense cloud of hydrogen that seemed not to have had a galactic origin and contained no stars, two features indicating that the gas may be primordial.

to be interacting with a pair of neighbors, the Large and Small Magellanic Clouds, creating the Magellanic Stream, a gaseous trail composed primarily of neutral hydrogen.

At the time of Zwicky's initial proposals for intergalactic activity, however, the astronomical community continued to be skeptical of the idea that matter of any sort could exist outside of galaxies. Support for the notion came indirectly, in 1960, when Jan Oort repeated the studies he had done in 1932 of stellar motions in the region around the Sun. In the decades since, the advent of more powerful telescopes had yielded data on stars previously too dim to see. But instead of eliminating the discrepancies he had found earlier, Oort came up with the same answer: About half of the mass in the Milky Way seemed to be missing.

This time Oort did not downplay the results. "This unknown population," he declared, "is the principal obstacle for deriving a model of the mass distribution of the galaxy." The statement was tantamount to sending up a warning flare. Because he was widely considered astronomy's greatest living practitioner, a sense of unease began to spread. Not only did individual galaxies like the Milky Way and clusters of galaxies like Virgo and Coma not contain enough luminous matter to explain their motions, but as scientists looked at larger and larger aggregations, they found that the discrepancies increased in direct proportion to the scale of the system.

AN INTERGALACTIC SEA

By the time Oort completed his recalculations, astronomers had expanded their reach by adding radio telescopes to their stellar arsenal and were gathering even more clues about missing matter. For one thing, they were learning that some of it actually did exist outside the conventional boundaries of galaxies visible at optical wavelengths—although the evidence for this unseen intergalactic medium was indirect.

In the early 1950s, observers discovered what are now known as radio galaxies, powerful emitters of radio waves whose optical counterparts often are utterly dwarfed by the extent of their radio emissions. The radio waves—at 408 megahertz, or just past very high television frequencies—appear to be given off by concentrations of extremely fast-moving plasma, subatomic particles such as electrons traveling near the speed of light. The plasma streams out tens of thousands of light-years to either side of the galaxy's optical core in narrow jets that lead to broad, gigantic radio lobes. Physicists have calculated that, over the course of its lifetime, whatever mechanism generates the jets would consume as much energy as would be released in the complete destruction of hundreds of thousands of stars such as the Sun. The leading candidate for the force at the heart of a radio galaxy is a supermassive black hole, which pulls gaseous material from its surroundings into a flattened accretion disk. As the disk rotates, it whips electromagnetic energy and particles outward at near light-speed, sending tightly focused beams in opposite directions along its spin axis.

THE CASE OF THE MISSING MATTER

One of the most intriguing puzzles facing astronomers today is that of so-called dark matter, or the missing mass problem. Starting in the early 1930s, scientists studying the velocities of stars in the Milky Way galaxy noticed that the mass necessary to account for their observed motions was twice as much as the mass the stars themselves seemed to embody.

The same discrepancy was seen in other individual galaxies—and also in galactic multitudes. In 1936, for example, astronomer Sinclair Smith of Mount Wilson Observatory examined the Virgo cluster *(below)*, an assemblage of 250 large star systems and at least a thousand lesser ones located about 70 million light-years from the Milky Way. Given their velocities and visible mass, the galaxies should be drifting ever farther apart. Instead—mystifyingly—they are held together by unseen mass 300 times greater than that of the luminous stars and dust.

White dwarfs. Stars roughly the size of the Sun end their lives as white dwarfs, retaining from 40 to 75 percent of their mass as they contract to about the size of Earth. These dim, dead stars thus meet the criteria for dark matter, but the universe is not old enough for a significant number of them to have formed.

Brown dwarfs. Lacking enough mass to ignite nuclear fires in their cores, these failed stars are one of the leading dark matter candidates. In theory, they range from slightly more than the mass of Jupiter to one-tenth the mass of the Sun, but no brown dwarfs have been positively identified as yet.

THE SEARCH FOR THE 90-PERCENT SOLUTION

Studies of galactic rotation rates in the early 1970s confirmed the disquieting fact that dark matter makes up at least 90 percent of the mass of the universe. In contradiction to Newton's and Kepler's laws of gravity and motion, rotation rates hold steady beyond the luminous edges of galaxies rather than falling off with distance from the center.

With galaxies apparently buried in invisible halos *(left)* that extend ten times the diameter of their luminous parts, the debate has focused on two theories of what this unseen material might be. One camp espouses the notion that it is ordinary but extremely dim matter such as failed stars. The other favors a host of exotic particles, many of them as yet purely theoretical. According to this hypothesis, as clouds of primordial gas cooled in the early universe, atoms of ordinary gas fell to the center to form galaxies, leaving halos of exotic particles behind. A gallery of candidates for dark matter is described below.

Neutrinos. Estimated to exist in numbers ranging from 10 to 100 million per cubic centimeter in galactic regions, neutrinos are among the most elusive particles detected; physicists have yet to determine their mass—if any. If they possess even a minute mass, they could account for the observed gravitational effects of dark matter.

Photinos. One of a class of exotic particles known as cold dark matter because they move more slowly than neutrinos, photinos are theoretically more massive than their hot counterparts but exist in far fewer numbers: A galactic halo would contain less than one per cubic centimeter—and none have been detected so far.

The fact that the narrow, high-speed jets suddenly spread out into more diffuse lobes suggested that this phenomenal outpouring of energy met with some sort of material resistance—an intergalactic medium that caused the matter in the jets to slow down abruptly. Similar evidence came some years later with the discovery of so-called head-tail galaxies, radio galaxies whose jets and lobes have a swept-back appearance, forming a kind of wake, as if the galaxies were plowing through an invisible ocean *(pages 104-105)*.

The nature of the cosmic sea between galaxies remained obscure, however. Given the lack of observed reddening or dimming of light from distant galaxies, scientists were fairly certain that the intergalactic medium, unlike the interstellar medium, held little or no dust. And even though they were beginning to find evidence for molecular gas in interstellar space, they were also convinced that the density of matter in the reaches between galaxies was far too low to allow molecules to form. Some astrophysicists suggested that intergalactic space might contain the last remnants of primordial gas, left over from the Big Bang, as well as ionized hydrogen that might have escaped as a result of galactic collisions. But detecting the spectral signatures of these gases—given their presumably tenuous distribution—would be difficult.

CREATION'S ECHO

Once again, however, the burgeoning field of radio astronomy wrested new clues from the sky. In 1964, Bell Telephone Labs' Arno Penzias and Robert Wilson (who six years later would discover carbon monoxide molecules in interstellar space) were struggling to set up a microwave antenna designed to communicate with orbiting satellites. But the sensitive receiver kept picking up radio interference, or noise. After systematically ruling out every earthly source, they concluded the noise had an extraterrestrial origin. As they aimed their dish toward various parts of the celestial sphere, they discovered that the signal they were catching was uniform rather than directional and that it represented energy whose temperature was only three degrees Kelvin above absolute zero.

This background radiation, as it is called, had been predicted some sixteen years earlier as a way to test the Big Bang theory of the origin of the universe. Because of the expansion of the universe, the reasoning went, the high-energy, short-wavelength radiation that permeated the fiery newborn cosmos at the moment of the Big Bang would have stretched, or red-shifted, until it was only a radio echo—low-energy, microwave radiation with a temperature of about five degrees Kelvin. The finding by Penzias and Wilson not only seemed to confirm the Big Bang theory but also sparked interest in intergalactic space as a region well worth investigating.

At about the same time the Bell Labs team was puzzling over extraneous radio noise, another group of scientists struggled to understand strange radio sources whose optical counterparts appeared to be dim stars—but stars whose optical spectra read as so much gibberish. Try as they might, scientists could discern no recognizable patterns of emission and absorption lines. The breakthrough, when it came, would present astronomers with a splendid new tool for probing the most distant reaches of the universe.

On February 5, 1963, Maarten Schmidt of Caltech sat studying a bit of film with the optical spectrum for one of these objects, designated 3C273. In a moment of inspiration, he suddenly saw a pattern: Three fuzzy lines at the far-red end of the spectrum, Schmidt realized, were actually three of four emission lines of hydrogen that normally appear toward the blue end. They had red-shifted so drastically that the fourth line had disappeared into the infrared. Schmidt calculated that 3C273 was receding at nearly 16 percent of the speed of light, which placed it almost three billion light-years away from Earth. Almost immediately, Schmidt's colleagues Jesse Greenstein and Tom Matthews solved the riddle of another quasar—short for "quasi-stellar radio source"—as the objects came to be known. (They are also designated QSOs, for "quasi-stellar objects.") This one, 3C48, was found to be receding at 37 percent of the speed of light, putting it at the then-known limits of the universe, or five to six billion light-years away. Scientists now theorize that these extraordinarily luminous objects, which radiate about a thousand times the energy of the entire Milky Way galaxy from an area scarcely the size of the Solar System, may be extremely massive black holes at the centers of galaxies.

Since quasars are not only powerful light sources but also extremely far away, they offer astronomers a way to investigate the voids between galaxies. In 1965, James Gunn and Bruce Peterson, then graduate students at Caltech, decided to use the light of quasars to search for intergalactic neutral hydrogen. Just as clouds of interstellar calcium and sodium produced so-called stationary absorption lines in the spectra of binary stars, Gunn and Peterson reasoned, so would intergalactic gas reveal its presence through absorption lines in the spectra of quasars.

CLUES FROM LYMAN-ALPHA

The key to their approach lay in a 1914 discovery by Harvard physicist Theodore Lyman, who showed that under certain conditions, neutral atomic hydrogen (HI) absorbed light at ultraviolet wavelengths. Working in the laboratory, Lyman produced a number of absorption lines, the first of which—alpha—occurred at 1,216 angstroms. He predicted that the line would be present in sunlight, but because Earth's atmosphere absorbs most ultraviolet wavelengths, he was unable to confirm his prediction. Lyman died in 1954, five years before rocket-borne instruments would rise above the atmosphere to record the Sun's Lyman-alpha line.

The advantage of quasars is that the enormous redshifts in their spectra resulting from their phenomenal recessional velocities bring the ultraviolet Lyman-alpha emission line into the visible spectrum, making it detectable by ground-based instruments. If neutral hydrogen was uniformly distributed in intergalactic space along the line of sight between Earth and a quasar, Gunn and Peterson theorized, it would produce broad absorption lines (forming a kind of trough) to the short-wavelength, or blue, side of the quasar's own Lyman-alpha emission line. The two researchers found no such trough. Instead, the quasar's spectrum was essentially flat on the short-wavelength side of the emission line. Gunn and Peterson concluded that intergalactic HI—assuming it was uniformly distributed—must have a density of less than one atom per 10 billion cubic centimeters, which was the limit of their instruments' sensitivity.

Meanwhile, others were also employing quasars to take core samples of the universe. In 1966, Margaret Burbidge and her husband, Geoffrey Burbidge, at the University of California at San Diego, searched the spectrum of the quasar 3C191 and found the paired absorption lines of magnesium on the blue side of the quasar's Lyman-alpha emission line. The Burbidges thought that the absorbing gas responsible for the lines was associated with the quasar, perhaps ejected by the brilliant object back along the line of sight to Earth. Three years later, however, John Bahcall and Lyman Spitzer of Princeton reviewed the data and argued that a better explanation for the lines was that they were produced by gaseous halos around otherwise unseen galaxies along the line of sight to the quasar. The two scientists based their argument on the fact that metals, as elements heavier than hydrogen and helium are known in the astrophysical community, are produced only in the nuclear furnaces of stars and could therefore exist only in close association with a galaxy. Scientists now believe both explanations are valid.

Whatever the case, evidence supporting the notion that some sort of gas exists between galaxies came in a 1971 paper by Roger Lynds of Kitt Peak National Observatory. The year before, Lynds was examining the spectrum of quasar 4C05.34, when he found a jumbled series of absorption lines, including five prominent Lyman-alpha lines at a variety of redshifts on the blue side of the quasar's own highly red-shifted Lyman-alpha line. He concluded that this Lyman-alpha forest, as such groupings are now known, was produced by many unevenly distributed, thin clouds of HI located in intergalactic space along the line of sight to the quasar. The distribution and exceedingly tenuous nature of the gas explained why Gunn and Peterson's equipment had not been able to detect it six years earlier. Lynds believed that the Lyman-alpha clouds were intergalactic, in part because they appeared to be made up only of atomic hydrogen, with no sign of metals that would suggest stellar (and, therefore, galactic) processes. This was of special note to astrophysicists because such concentrations of HI would presumably be leftover material from the Big Bang.

A NEW WINDOW

Although radio and optical astronomers have not detected direct evidence of a diffuse and continuous intergalactic medium, exploration of the regions between galaxies continues at other windows in the electromagnetic spectrum—notably one that opened just before midnight on June 18, 1962. On that date, a group of researchers from Cambridge-based American Science and

PROBING DEEP SPACE WITH A QUASAR'S LIGHT

The spectra of quasars billions of light-years away have become powerful tools for exploring the depths of intergalactic space. Because a quasar is so distant, its light is extremely red-shifted by the time it reaches Earth. Thus, very short wavelength ultraviolet light, which is normally blocked by Earth's atmosphere, is lengthened into the visible range, yielding spectral lines that would not otherwise be detectable by ground-based instruments.

In deciphering quasar spectra *(below)*, scientists start with a telltale marker: a so-called Lyman-alpha emission line, produced by neutral atomic hydrogen, that ordinarily occurs in the ultraviolet. The degree to which this line has shifted into the visible tells scientists how far away the quasar is.

On the blue, or short-wavelength, side of this emission line are myriad dark absorption lines whose redshifts are less than that of the quasar's Lyman-alpha emission line, indicating that the objects producing them lie between Earth and the quasar. One group of lines, known as the Lyman-alpha forest, is generated when the quasar's light passes through successive clouds of cool intergalactic hydrogen; different distances result in different redshifts, hence the "forest." Another consists of absorption lines produced by such heavy elements as carbon, silicon, and magnesium. Since these metals, as they are called, are created by stellar processes, scientists believe the absorptions occur when quasar light slices through the gaseous halos of otherwise invisible galaxies.

When ultraviolet light from a quasar passes through a series of objects *(left)*, the resulting spectrum contains a welter of emission and absorption lines that show varying degrees of redshift as compared to the quasar's theoretical at-rest spectrum. The individual spectra below, broken out of the overall quasar spectrum, reveal the presence of intergalactic gas clouds and galaxies at different distances along the line of sight from Earth.

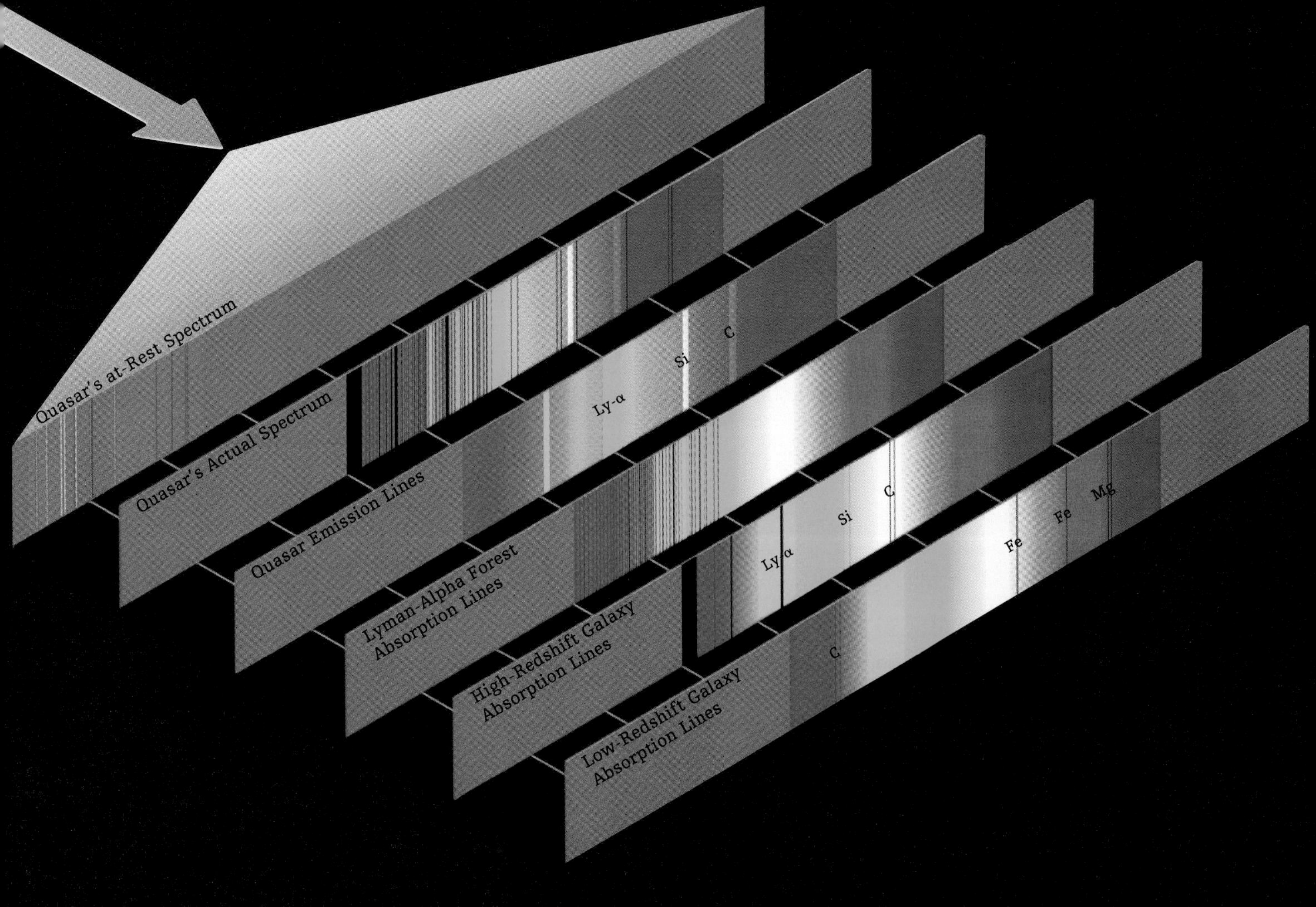

Quasar's at-Rest Spectrum

Quasar's Actual Spectrum

Quasar Emission Lines

Ly-α Si C

Lyman-Alpha Forest Absorption Lines

High-Redshift Galaxy Absorption Lines

Ly-α Si C Fe Fe Mg

Low-Redshift Galaxy Absorption Lines

C

Engineering (AS & E) launched a sounding rocket that carried three x-ray detectors into Earth's upper atmosphere to make two spectacular discoveries.

First, the detectors located a discrete source of x-rays in the constellation Scorpius, designated Sco X-1, that was much more intense than scientists had expected from even the nearest normal star. Subsequent studies over the years have revealed several hundred such x-ray emitters, most of them binary systems in which one companion is a very small but massive star—a white dwarf, a neutron star, or, perhaps, a black hole.

An even more unexpected finding came as the rocket spun around to sweep the celestial sphere: In addition to Sco X-1, the detectors picked up a consistent reading of x-radiation in all directions. Astronomers realized that this diffuse background radiation could not be coming from sources within the galaxy or even from nearby galaxies. Rather, it must come from so far away that emanations from all discrete sources simply merged, forming a smooth background. Like the three-degree microwave background that would be discovered two years later, the x-ray background held critical information about the origin and structure of the universe.

Just what the source of this background x-radiation was remained a puzzle, however, and further advances were long in coming. The next major effort in the field of x-ray astronomy occurred nearly a decade later. On December 12, 1970, NASA launched *Uhuru*, a satellite that was designed to conduct the first extended x-ray survey of the sky *(right)*. Among the project members who awaited the craft's findings was Herbert Gursky, a member of the AS & E team that had worked on the x-ray rocket eight years earlier. Upon examining *Uhuru*'s results, Gursky and his colleagues found that twenty x-ray sources could be matched with known clusters of galaxies, including Perseus, Virgo, and Coma, the cluster whose excess mass had bewildered Fritz Zwicky in 1933. The radiation seemed to come from each cluster as a whole, rather than from individual galaxies—a finding that strongly bolstered Zwicky's argument that galaxy clusters contain intergalactic matter not visible at optical wavelengths.

A host of follow-up satellites helped fill in the picture of the x-ray universe. Spectrographic results that were relayed in 1976 by *Ariel 5* (fifth in the British Ariel series launched between 1962 and 1979), for example, led Richard J. Mitchell of the Mullard Space Science Laboratory of University College, London to conclude that atoms of highly ionized iron—in which twenty-four of the atom's ordinary complement of twenty-six electrons have been stripped away—exist within the Perseus cluster of galaxies. At the same time, Peter J. Serlemitsos and colleagues at the NASA Goddard Space Flight Center in Greenbelt, Maryland, looked at the data from a satellite designated *OSO-8* and found similar x-ray signatures from highly ionized iron inside the Virgo and Coma clusters.

All of these studies suggested something significant about the intergalactic medium: The gas between galaxies that are gravitationally bound in clusters is not primordial. Finding heavy elements in intracluster gas indicates that

Blasting off from a modified oil platform near Kenya's coast, a NASA rocket carries the first x-ray satellite into space on December 12, 1970—Kenya's independence day. The launch date inspired the satellite's nickname, *Uhuru*, which means "freedom" in Swahili. *Uhuru* picked up x-ray readings that led to the discovery of emission zones within clusters of galaxies. Once considered empty, these intergalactic voids are now thought to be filled with low-density hot gas at temperatures around 100 million degrees.

the gas somehow escapes from galaxies *(pages 125-135)*. For example, it might be dragged out by gravitational forces resulting from near misses between galaxies, as Zwicky had long ago proposed. Or exploding supernovae might heat the interstellar gas until it becomes a kind of wind that escapes the grip of the galaxy's gravity.

For any given cluster of galaxies, according to Herbert Gursky, the nature of the cluster's intergalactic medium will vary depending on the evolution of the stars in the member galaxies. "Looking back," he has said, "we should see clusters with different amounts of gas in them."

Still unresolved is the source or sources responsible for the diffuse x-ray background. Most scientists agree that the process of galaxy formation in the early universe was probably less than completely efficient and that therefore a certain amount of primordial gas could have been left over outside of existing galaxies. However, any such gas is unlikely to be hot enough to account for the x-ray background, whose spectrum suggests a source at temperatures of at least 400 million degrees Kelvin.

In any case, heating gas to this temperature is problematic. Even quasars would have to put out considerably more energy than they do to generate the requisite heat. Moreover, such a hot background would evaporate most of the gas from galaxy clusters, which clearly has not happened. Indeed, for a variety of reasons, some astronomers theorize that quasars themselves, since they are so vastly distant, are the source of the x-ray background. These researchers expect that improved x-ray observatories in space will bring ever more quasars to light, effectively resolving the diffuse background into discrete sources.

BEYOND CLUSTERS

Meanwhile, although x-ray studies offer direct evidence of gas within galaxy clusters, so far they have not answered the question of whether such gas exists between clusters. More productive investigations along those lines have continued to come through the use of quasar spectra as probes.

In 1980, Wallace Sargent at the California Institute of Technology, along with colleagues at University College, London and at Caltech, started a systematic examination of the Lyman-alpha forests in five quasar spectra to determine their general features as well as the density of the clouds of neutral hydrogen distributed along the line of sight between quasars and Earth. By examining the absorption lines, Sargent could estimate the number of absorbing atoms along the line of sight, and he arrived at a density range of between 100 and 4,000 atoms per cubic meter.

Since this is considerably too tenuous to block the ultraviolet background radiation that permeated space in the distant era when quasars were plentiful, Sargent realized that most of the hydrogen in these intervening Lyman-alpha clouds must in fact be ionized. In other words, the neutral hydrogen detectable by Lyman-alpha absorption lines represented only one one-hundredth of a percent of the hydrogen in the clouds.

Sargent's work and subsequent investigations would show that these intercluster clouds are as large as small galaxies—spanning on the order of a few tens of thousands of light-years and containing the mass of 10 million to 100 million Suns. Estimates of their temperature suggest that the clouds are relatively warm—about 30,000 degrees Kelvin. Finally, these studies show that many of these clouds show no evidence of being contaminated by heavier elements, which means they are almost certainly primordial in origin and occupy the reaches outside galaxy clusters. (Scientists assume helium is also present in the clouds if the clouds are primordial, but helium absorption lines would occur too far in the ultraviolet region of the spectrum to be detected from Earth.)

PRIMORDIAL MATTER?

By the early 1980s, astronomers had sorted out a number of issues regarding the existence of matter outside of galaxies. X-ray observations of galaxy clusters and the detection of metal absorption lines in quasar spectra supported claims for gas in intergalactic space within clusters or otherwise associated with intervening galaxies. The discovery of the Lyman-alpha forest in quasar spectra spoke to the distribution of primordial hydrogen in the regions outside clusters. But all of these were in some sense indirect signals from deep space. Then, in late December of 1982, a more direct clue arrived.

Stephen E. Schneider, then a twenty-five-year-old student working on his astronomy doctorate at Cornell University, had traveled to Arecibo, Puerto Rico, for a session on the world's biggest radio telescope, operated by the university for the National Science Foundation. On December 28, he used the giant 1,000-foot dish to examine a supposedly empty portion of the sky in the constellation Leo. He wanted to adjust the telescope to ensure that its measurements would be accurate—a process called calibrating. To do that, the telescope had to be looking toward a part of the sky that did not contain radio sources. But the blackness refused to cooperate; he kept getting a signal.

Schneider was confused. "This happened on my very first run at Arecibo," he recalled later. "I was not sure what was going on. It was very exciting." At first he thought it might be interference from stray radar signals, for example. But on January 2, he looked toward Leo again and found that the signal was still there. Schneider then checked the Palomar Sky Survey, a collection of photographic plates of the northern sky, but the survey showed no galaxies at that position.

Schneider had detected 21-centimeter radio signals from a gas cloud between the galaxies in the direction of the constellation Leo. The cloud, Schneider noted, was huge, spanning twice the area of the full Moon in the sky, even though it was 30 million light-years away. Moreover, the gas was not attached to a galaxy. Detailed mapping of the area around the gas cloud in both optical and radio frequencies failed to find any galactic sources for the cloud, eliminating the possibility of a recent gravitational war between two galaxies that

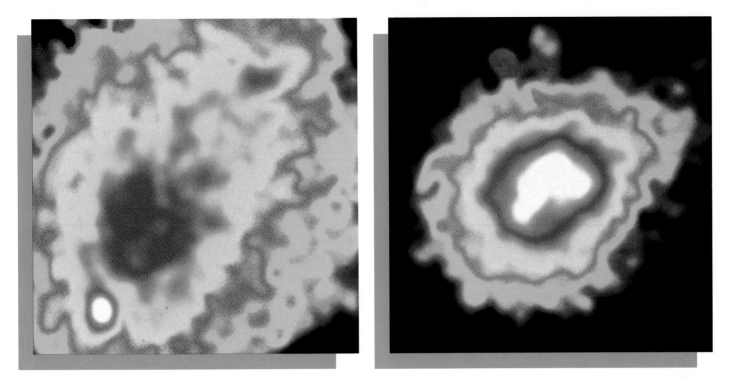

X-rays from distant clusters of galaxies reveal clues about the age and structure of the universe. The coded colors reflect x-ray intensity—ranging from the hottest and brightest in white, through red and yellow, to the coolest and dimmest in blue. The irregular shape, low luminosity, and low temperature of the cluster A1367 *(above)* suggest that the cluster is relatively young as compared to A2256 *(above, right)*, a very hot cluster whose rough edges have been smoothed over time as its galaxies shifted inward to create a hot, bright core. Scientists infer from these patterns that galaxies formed before clusters, gathering first in clumpy groups and later sorting themselves into more symmetrical arrangements filled with hot gas.

might have sucked substance from the weaker of the two. Finally, because the cloud emits 21-centimeter radio waves and shows other signs of being made up of cool gas, Schneider theorized it must contain mostly neutral atomic hydrogen—in which case it may well be primordial in origin.

Such huge intergalactic clouds may yet prove to be the birthplaces of galaxies. In 1986, Arthur Wolfe, then at the University of Pittsburgh and later at the University of California at San Diego, along with colleagues from both schools and from the Space Telescope Science Institute in Baltimore, completed a survey of sixty-eight quasar spectra. Based on certain characteristics in the Lyman-alpha signatures, the group identified what they believe are more than a dozen protogalaxies along the lines of sight to the various quasars. For example, the Lyman-alpha absorption lines of these otherwise invisible objects were noticeably broader than those produced by typical Lyman-alpha clouds, suggesting that the newly discovered objects had a high density of hydrogen along the line of sight. In addition, data from radio telescopes indicated that the objects were disk shaped and had temperatures of only a few hundred degrees Kelvin. Further research on one disk revealed Lyman-alpha emission lines, an indication that the cloud may contain stars and may, in the end, prove to be a stellar nursery.

AN ENDURING RIDDLE

Despite the excitement of discovering matter in the black regions between stars and galaxies once held to be utterly void, researchers are no closer to finding the answers to the missing mass question. Indeed, the discrepancy between the motions of stars and galaxies and their observable mass took a turn for the worse in the late 1960s.

Starting in 1969 with Vera Rubin and Kent Ford, both at the Carnegie

Institution of Washington, and continuing a few years later with Jeremiah Ostriker and James Peebles, two theorists at Princeton University, astronomers have learned that galaxies seem to be embedded in halos of matter undetectable except by their gravitational effects. Rather than rotating ever more slowly with distance from the galactic center, stars at the outer edges of galaxies do not slow down as predicted but maintain their speed, as if there is more mass beyond them. Ostriker and Peebles developed a computer model to test the notion that the mass of spiral galaxies would be distributed in a pattern similar to their light: denser at the center than toward the edges. The Princeton model showed that a galaxy with a distribution of diminishing mass would be unstable and would fly apart. The only way the researchers could get the model galaxy to stabilize was to surround it with a massive halo of unknown origin.

Cosmologists have come up with a number of possible explanations about the form that undetectable material might take *(pages 113-115)* ranging from various types of dead or failed stars to entirely theoretical objects, such as monopoles, gravitinos, and cosmic strings. As technological and theoretical advances bring more and more corners of the universe to light, the question of missing mass grows increasingly frustrating, prompting some physicists to propose that a breakdown in Newtonian gravity itself may somehow be at fault. Certainly Fritz Zwicky would have relished the opportunity to gloat over colleagues who had dismissed his concerns about this issue. If nothing else, however, in the decades since he challenged the status quo, astronomers have succeeded in penetrating the cosmic darkness to pry out at least some of the secrets of matter hidden in the regions between the stars.

eep space, at first glance, appears dark, empty, and cold. Earthbound optical and radio telescopes detect gas and dust between stars, but even the best such instruments discern no sign of matter in the gaps between galaxies. Yet those dark recesses turn out to be neither empty nor cold. In 1971, a satellite orbiting beyond Earth's atmospheric x-ray filter provided the first evidence of intergalactic matter: gas up to ten times as hot as the fuel at the Sun's core, so torrid it emits nothing but x-rays.

Puzzling over this discovery, astronomers concluded that much of the gas is left over from the era of galaxy formation. As galaxies condensed out of clouds of gas, they gleaned only a portion of the available material, leaving vast amounts behind. But this residue cannot account for all intergalactic matter detected from x-rays. For every million atoms of hydrogen, surveys reveal about ten atoms of iron and small concentrations of other heavy elements, which can only be forged within stars. Once they are discharged by the stellar explosions known as supernovae, they must somehow escape their galaxy's gravitational field to pollute the primordial hydrogen beyond.

Such a flight requires an enormous amount of energy. To break the grip of galactic gravity, the gas must reach at least 10 million degrees Kelvin. At that temperature, the individual atoms move randomly at a velocity of 300 miles per second, and the gas expands with enough force to sever the galactic tie. Since this process unfolds over huge stretches of time and space, it has largely eluded observation. But astronomers theorize that it might occur in several ways, depending on the type of galaxy involved and on its position within a galaxy cluster. As illustrated on the following pages, galaxies may shed gas when they collide, when they are kindled internally by fusillades of supernovae, when they orbit through hot zones within their cluster, or when they whip through the denser gas near the cluster's core.

At near right, two spiral galaxies approach each other nearly head-on. In the next frame, some 450 million years later, they emerge from their collision trailing long tails sculpted by gravitational interactions. After another 150 million years *(far right),* the galaxies have receded from each other, leaving behind a vast amount of gas freed by the shock of the encounter. In the final view *(below),* roughly 400 million years after the collision, escaped gas pervades the gap between the galaxies.

THE SHOCK OF GALACTIC COLLISIONS

The term galactic collision suggests a sudden and spectacular cataclysm, capable of blowing stars to bits, pulverizing planets, and liberating galactic gas along the way. In fact, the process is excruciatingly slow, played out over hundreds of millions of years; during that time, powerful gravitational tides may pull stars from their galactic moorings, but the stars within colliding galaxies are so widely spaced that

they stand virtually no chance of coming in contact.

On a microcosmic level, however, galactic encounters may be violent indeed. The clouds of gas within each galaxy are composed of atoms that lie relatively close together compared to stars—too close to allow the clouds to interpenetrate peacefully. When two gasrich spiral galaxies bear down on each other with enough speed, as illustrated at top, the swarming at-

oms clash epochally. The encounter produces great heat, agitating atoms until they break the clutch of galactic gravity. As the impact knocks gas out of the galaxy's orbital flow, the gas begins to lag behind the galaxy that once embraced it.

The combined effect on the galaxies of this impact and the subtler gravitational tides is detailed above. In the wake of the encounter, tidal forces endow each

galaxy with a long tail that curls toward the opposing galaxy. Most of the material within these tails is still bound by gravity. But in between the receding galaxies lies a zone of hot gas liberated by the shock of the collision. Free of galactic gravity, this hot gas diffuses so widely that its individual particles seldom come close enough to each other to release energy in the form of light and cool off.

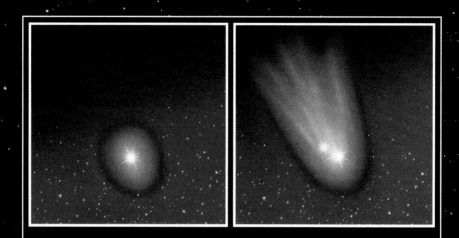

Above, at left, a superbubble of gas *(orange)* heated by supernovae to a temperature of several million degrees expands against the pressure of the cooler, denser interstellar medium *(purple)*. As the bubble nears the edge of the galaxy, it gains impetus from additional supernovae and meets with less resistance from the thinning galactic medium. Finally, several million years after its genesis, the superbubble bursts through *(above, right)*, ejecting hot gas from the galaxy to enrich the intergalactic medium.

SUPERNOVA VENTS IN SPIRAL GALAXIES

In a spiral galaxy, multiple supernovae can generate enormous shock waves that heat and accelerate the cool, dense gas between stars, literally blowing it out of the galaxy. A single supernova may heat the surrounding gas atoms by more than a million degrees, forming a bubble that expands for hundreds of thousands of years before diminishing. One such supernova remnant is not hot enough to liberate gas from the galaxy; but within a single region of some spiral galaxies, tens or even hundreds of stars may reach the same stage in their development and detonate in sequence, inflating the bubble into a superbubble.

As the superbubble expands over the eons, it approaches the edge of the galaxy *(box, left)*. There the confining pressure of cool, dense galactic gas lessens and the upward momentum of the shock wave increas-

es, stretching the spherical superbubble into an egg shape. Given a sufficient boost from additional supernovae, the pocket of seething gas will ultimately break out of the galaxy. Gas atoms with enough energy soar away *(above),* free forever of galactic gravity. Particles with somewhat less energy will succumb to the gravitational pull and fall back toward the galaxy.

The more supernovae that contribute to the super-bubble, the stronger the chimney, as such geysers of hot gas are known. Hundreds of chimneys at a time may shoot hot gas out of a large spiral galaxy, carrying away the equivalent in mass of a thousand Suns in as many years. Among the particles ejected in the process are telltale traces of iron and other star-born elements whose x-ray signals first alerted astronomers to the galactic origins of intergalactic matter.

Supernova Winds in Elliptical Galaxies

Elliptical galaxies differ from spiral galaxies not only in shape but in evolution and composition, and the differences affect the way supernovae kindle galactic gas to the escape point. Astronomers believe that elliptical galaxies result from a highly efficient process of star formation—punctuated by frequent supernovae—that leaves little primordial gas behind. (Spiral galaxies, by contrast, have plenty of original gas left

over, and barring an infusion of kinetic energy from supernovae or other sources, the gas remains relatively dense and cool.)

Since the gas in a mature elliptical galaxy is sparse and hot, there is no mantle of cool, concentrated particles to contain the shock wave of a supernova. Instead, the wave spreads evenly through the meager interstellar medium, heating up the gas uniformly by

millions of degrees. Relatively few supernovae are required to raise most of the gas in the galaxy to the escape temperature. Unlike the chimney effect associated with spiral galaxies, the torrid gas in an elliptical disperses from the entire surface of the galaxy —a phenomenon known as galactic wind.

The one thing that prevents such a wind from dispelling all the gas in the galaxy is the diminishing supply of supernovae in an elliptical. Compared with spiral galaxies, ellipticals are old and settled, their youngest stars thought to be more ancient than the Sun. Having already passed through their volatile formative phase, these veteran galaxies experience fewer supernovae as they age—enough to keep them relatively gas-poor without necessarily exhausting their interstellar matter.

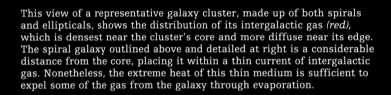

This view of a representative galaxy cluster, made up of both spirals and ellipticals, shows the distribution of its intergalactic gas *(red)*, which is densest near the cluster's core and more diffuse near its edge. The spiral galaxy outlined above and detailed at right is a considerable distance from the core, placing it within a thin current of intergalactic gas. Nonetheless, the extreme heat of this thin medium is sufficient to expel some of the gas from the galaxy through evaporation.

EVAPORATION: GALAXIES AT A BROIL

A galaxy orbiting within a cluster glides through a rarefied cloud of hot intergalactic gas that is loosely bound by the cluster's gravity. According to one theory, this searing, ambient gas can cook the orbiting galaxy until its own gas is agitated to the escape point and expelled like the air in an overheated soufflé.

This process—referred to by astronomers as evaporation—would operate efficiently only under certain conditions. For one thing, the intergalactic magnetic field lines must be straight. Intergalactic gas is ionized, conducting heat to the orbiting galaxy through a flow of electrons; if the field lines are tangled, that flow will be impeded, thwarting heat conduction.

Assuming that the field lines are straight, the gas enveloping a galaxy would have to be both extremely hot and relatively dense to produce much evaporation.

These optimum conditions are not pervasive, since the gas cools as its density increases toward the gravitational core of a cluster. At the outer edges of a cluster, intergalactic gas is extraordinarily hot, running as high as 100 million degrees. But its density is correspondingly low, and the heat is so diffuse that galaxies orbiting there experience little evaporation. Near the core, the density of the intergalactic gas is much great-

er, but its temperature is significantly lower. The best conditions for evaporation lie between these two extremes—roughly midway between the core of the cluster and its edge. In the example illustrated above, a spiral galaxy is orbiting beyond the median point, about a third of the way from the edge. There the intergalactic gas is quite hot but not very dense, resulting in a moderate amount of evaporation.

Ram Pressure: Paring a Galaxy Down

Although the dense gas near the core of a galaxy cluster may cause little evaporation, it can strip a passing galaxy of much of its interstellar matter through a mechanism called ram pressure. In earthly terms, ram pressure is exemplified by the bicyclist who whizzes downhill on a calm day, pushing through billions of air molecules and perceiving the pressure as a stiff wind. The force of this wind is a function of the speed of the cyclist and the density of the opposing air.

On a cosmic scale, speed and density conspire to create enormous ram pressure when a galaxy tracing an elongated orbit within a large cluster nears its core, as illustrated above. Approaching the core, the galaxy accelerates in response to the increasing tug of gravity; at the same time, the opposing medium of intergalactic gas grows denser. The combined effect is to magnify ram pressure up to a thousand times. Here, as in collisions between galaxies, the clash of opposing masses of atoms heats the orbiting galaxy's gas and knocks it off course, causing it to be left behind as

the galaxy swings back toward the cluster's edge.

Galactic gas liberated in this way may enjoy only partial freedom. Many clusters contain a central, giant elliptical galaxy—represented above as a pincushion of light—which exerts a strong gravitational hold on nearby intergalactic gas. Even if there is no such galaxy, each cluster possesses a center of gravity that keeps the surrounding gas fairly dense.

The impoverished orbiting galaxy may eventually replenish its interstellar matter as it travels through the far reaches of the cluster, where ram pressure is reduced. Even gas-poor elliptical galaxies, such as the one orbiting here, can increase their meager store through stellar evolution; some dying stars, for instance, bequeath their contents to the galaxy with far less fanfare than do supernovae, giving off plenty of gas without as much depleting heat. Yet any gains made by the galaxy are temporary. Eventually its orbital momentum carries it back toward the cluster's dense hub, where ram pressure again takes its toll.

GLOSSARY

Absorption line: a dark line or band at a particular wavelength on a spectrum, formed when a substance between a radiating source and an observer absorbs electromagnetic radiation of that wavelength. Different substances produce characteristic patterns of absorption lines.

Ammonia: a molecule made up of one atom of nitrogen and three of hydrogen. Astronomers use ammonia to gauge the density and temperature of interstellar clouds because its unique energy state transitions are readily observable at microwave wavelengths.

Angstrom: a unit of length equal to one ten-billionth of a meter (about four-billionths of an inch); used in astronomy as a measure of wavelength.

Asteroid: a small, rocky, airless body that orbits a star.

Astronomical unit (AU): a unit of measure, often used to express distances within the Solar System, that is equal to the mean distance between Earth and the Sun, or 92,960,116 miles.

Binary stars: a gravitationally bound pair of stars in orbit around their mutual center of gravity. Binary stars are extremely common, as are systems of three or more stars.

Black hole: theoretically, an extremely compact body with such great gravitational force that no radiation can escape from it.

Blueshift: a Doppler effect seen when a radiating source approaches the observer. The received wavelengths shorten so that any absorption or emission lines move from their expected frequencies toward the short-wavelength, blue end of the spectrum. *See* Redshift.

B stars: a spectral class of young, massive, blue stars, the second brightest type, found mainly in the galaxy's spiral arms. Because they emit high levels of radiation, B stars ionize the hydrogen in surrounding clouds.

Calcium: a chemical element common in its solid state on Earth and present as a gas in interstellar space.

Carbon monoxide: a diatomic compound of carbon and oxygen found on Earth as a toxic by-product of combustion and present in large quantities in interstellar space.

Cepheid variable: a star that changes regularly in luminosity over a set period of days or weeks.

Cluster: a gravitationally bound system of galaxies, ranging in number from a few dozen to several thousand.

Constellation: originally a pattern of stars named for an object, animal, or person, but now more commonly the area of sky assigned to that pattern. Every astronomical object is located in a specific constellation.

Cosmic ray: an atomic nucleus or other charged particle moving at close to the speed of light; thought to originate in supernovae and other violent celestial phenomena.

Dark matter: a form of matter that has not yet been directly observed, but whose existence is deduced from its gravitational effects.

Diatomic: having two atoms in a molecule.

Doppler shift: a change in the wavelength and frequency of sound or electromagnetic radiation, caused by the motion of the emitter, the observer, or both.

Dust: small grains of matter, largely graphite and silicates, that pervade interstellar space. Although extremely diffuse, dust is responsible for scattering, absorbing, and polarizing starlight. Dust grains are also the building blocks of most celestial matter and the site of chemical reactions that produce interstellar molecules.

Electromagnetic radiation: waves of electrical and magnetic energy that travel through space at the speed of light. *See* Electromagnetic spectrum.

Electromagnetic spectrum: the array, in order of frequency or wavelength, of electromagnetic radiation, from low-frequency, long-wavelength radio waves, through infrared, visible light, and ultraviolet, to high-frequency, short-wavelength gamma rays.

Electron: a negatively charged particle that normally orbits an atom's nucleus but may exist in isolation.

Emission line: a bright band at a particular wavelength on a spectrum, emitted directly by the source and indicating by its wavelength a chemical constituent of that source.

Energy level: a value of energy associated with electrons in an atom or with the spin or vibration of a molecule. The lowest possible energy level is called the ground state; higher levels, brought about by an increase in energy, are called excited states.

Extinction: in astronomy, the blocking of light from stars or other celestial bodies caused by a combination of absorption and scattering.

Formaldehyde: an organic compound of hydrogen, carbon, and oxygen, found both on Earth and in space.

Frequency: the number of oscillations per second of an electromagnetic (or other) wave. *See* Wavelength.

Fullerene: in theory, a large carbon-based molecule thought to occur in soot on Earth and in space.

Galactic plane: the central plane of the Milky Way galaxy; also, the central plane of any disk-shaped galaxy.

Galactic rotation: the movement of matter—including stars, gas, and dust—around the center of a galaxy.

Galaxy: a system that contains from millions to hundreds of billions of stars as well as varying quantities of gas and dust.

Gamma ray: the most energetic form of electromagnetic radiation, with the highest frequency and the shortest wavelength.

Gas: matter in its most diffuse state, having neither a definite shape nor volume. Gas accounts for a significant fraction of a galaxy's mass. Most gas in the universe is a form of hydrogen.

Globule: a dense concentration of interstellar gas and dust, thought to be a protostar in the process of formation.

Gravity: a fundamental force of nature, the mutual attraction of separate masses.

Helium: the second lightest chemical element and the second most abundant.

Hydrogen: the most common element in the universe. It occurs in several forms, including neutral hydrogen (HI), hydrogen atoms, composed of one proton and one electron, with no electrical charge; ionized hydrogen (HII), positively charged hydrogen atoms from which the electron has been stripped; and molecular hydrogen, two-atom molecules of hydrogen.

Hydroxyl: a highly reactive, two-atom molecule of hydrogen and oxygen; in interstellar space, it often operates as a maser.

Infrared: a band of electromagnetic radiation with a lower frequency and longer wavelength than visible light.

Interferometer: in radio astronomy, an arrangement of two or more separated radio telescopes used as one to receive sources of radio emission. By studying the overlapping wave patterns produced from the emissions, scientists can determine the brightness and structure of the emitting source.

Interstellar medium: clouds of dust and gas between the stars.

Inverse-square law: the mathematical relationship that describes the change in brightness of a star, or other point source of light, that occurs in inverse proportion to the square of the distance from the source; also any similar mathematical formula that describes how certain forces, such as gravity, change in strength with distance from a central point.

Ion: an atom that has lost or gained one or more electrons, changing its electrical charge. In comparison, a neutral atom has an equal number of electrons and protons, giving the atom a zero net electrical charge. A positive ion has fewer electrons than the neutral atom; a negative ion has more.

Kelvin: an absolute temperature scale that uses Celsius degrees but sets 0 at absolute zero, or about -273 degrees Celsius.

Kinetic energy: an object's energy of motion.

Light-year: an astronomical unit of distance equal to the distance light travels in a vacuum in one year, or almost six trillion miles.

Lyman-alpha line: a strong spectral line produced by hydrogen at a wavelength of 1,216 angstroms.

Magnetic field: a region within which the effects of magnetism are observable. The strength of the field is indicated by the force exerted on moving charged particles within it.

Magnitude: a designation of an object's brightness or luminosity relative to that of other objects. Apparent magnitude refers to observed brightness; absolute magnitude refers to an object's hypothetical brightness at a standard distance of about 32.6 light-years from the observer.

Maser: a source of radiation that produces intense radio beams; hydroxyl molecules and water molecules have been found to act as masers.

Mass: a measure of the total amount of material in an object, determined either by its gravity or by its tendency to stay in motion if in motion, or at rest if at rest.

Mass-luminosity (M-L) relation: a formula relating the mass of a star to its luminosity; in general, the larger the mass, the brighter the star.

Metastable state: an intermediate energy state in an atom, higher than the lowest, or ground, state but lower than excited states.

Methylidyne: a two-atom molecule of carbon and hydrogen; the first molecule discovered in interstellar space.

Microwave: a radio wave of very high frequency and short wavelength.

Milky Way: the Earth's galaxy, a giant spiral of at least 100 billion stars. The Sun is two-thirds of the way out from the Milky Way's center.

Molecular cloud: a concentration of interstellar gas and dust sometimes hundreds of light-years in diameter. Such clouds are the most massive objects in the Milky Way.

Molecule: the smallest unit of an element or compound that retains its properties. A molecule commonly consists of two or more atoms bonded together.

Nebula: a cloud of interstellar gas or dust or a mixture of both. Nebulae may glow with their own or reflected light, or they may obscure light from other sources, producing dark patches in the night sky.

Neutrino: a chargeless subatomic particle with little or no mass.

Neutron: an uncharged particle with a mass similar to that of a proton; normally found in an atom's nucleus.

Nucleation: the start of the process of condensation, during which a core particle acts as a nucleus around which molecules collect.

O stars: a spectral class of massive, young, blue stars, the most luminous type, found primarily in the spiral arms of the galaxy. Because they emit high levels of radiation, O stars ionize the hydrogen in surrounding clouds.

Parsec: an astronomical unit of distance equal to approximately 3.26 light-years.

Period-luminosity law: a relation linking a Cepheid variable's absolute magnitude to the length of its period, or cycle of brightening and dimming (the longer the period, the brighter the star); a critical tool for estimating intragalactic distances. By comparing a Cepheid's absolute magnitude with its apparent magnitude, astronomers can calculate the star's distance.

Photon: a unit of electromagnetic energy associated with a specific wavelength. It behaves as a chargeless particle traveling at the speed of light.

Polarization: the tendency of some electromagnetic waves to vibrate preferentially in a single plane rather than uniformly in all directions perpendicular to their motion. Polarization is produced by the source of the radiation and by the medium through which it travels.

Polycyclic aromatic hydrocarbon (PAH): a complex, carbon-based molecule with a ring structure, found in soot on Earth. PAHs are thought by some scientists to be present in large quantities in interstellar clouds.

Proton: a positively charged particle, normally found in an atom's nucleus, with 1,836 times the mass of an electron.

Pulsar: a radiating source from which bursts of energy are received at precisely spaced intervals of several seconds or less. Pulsars are thought to be rapidly rotating neutron stars with very strong magnetic fields.

Quantum mechanics: a mathematical description of the rules by which subatomic particles interact, decay, and form atomic or nuclear objects.

Quantum tunneling: a quantum effect invoked to explain the movement of subatomic particles through otherwise impenetrable force barriers, such as into or out of an atomic nucleus.

Quasar: shortened from "quasi-stellar radio source"; an extremely powerful, bright source of energy, located in a very small region at the center of a distant galaxy, that outshines the whole galaxy around it.

Radial velocity: the speed of a celestial body approaching or receding in an observer's line of sight, estimated from Doppler shifts in the body's spectrum.

Radio: the least energetic form of electromagnetic radiation, with the lowest frequency and the longest wavelength.

Radio telescope: an instrument for studying astronomical objects at radio wavelengths.

Redshift: a Doppler effect seen when a radiating source recedes from the observer. The received wavelengths lengthen so that any absorption and emission lines move from their expected frequencies toward the long-wavelength, red end of the spectrum. *See* Blueshift.

Rotation: the turning of a celestial body about its axis.

Spectral line: *see* Absorption line; Emission line.

Spectrograph: an instrument that splits light or other electromagnetic radiation into its individual wavelengths—producing a spectrum—and records the result.

Spectroscopy: the study of spectra, including the position and intensity of emission and absorption lines.

Spectrum: the array of electromagnetic radiation, arranged in order of wavelength, from long-wave radio to short-wave gamma rays. Also, a narrower band of wavelengths, as the visible spectrum, in which light dispersed by a prism or other means shows its component colors, often banded with absorption or emission lines. A continuous spectrum is one that includes all wavelengths or frequencies, such as that produced by white light.

Stationary line: in stellar spectra, an absorption line that exhibits a Doppler shift different from other lines in the same spectrum, indicating that the object producing the line intervenes along the line of sight between Earth and the background stars. Stationary lines of calcium and sodium were the first evidence of interstellar gas.

Supernova: a stellar explosion that expels all or most of the star's mass and is extremely luminous.

21-centimeter line: a spectral line produced by neutral hydrogen at a radio wavelength of just over 21 centimeters; the first radio spectral line to be detected.

Ultraviolet: a band of electromagnetic radiation with a higher frequency and shorter wavelength than visible blue light.

Violet: radiation appearing at the extreme blue end of the visible spectrum, having a short wavelength and high frequency.

Wavelength: the distance from crest to crest, or trough to trough, of an electromagnetic or other wave. Wavelengths are related to frequency: The longer the wavelength, the lower the frequency.

X-radiation (x-ray): radiation intermediate in wavelength between ultraviolet and gamma rays.

Zeeman effect: the splitting of a single spectral line into two or more component lines, caused by the action of a magnetic field.

BIBLIOGRAPHY

Books

Abbott, David (ed.). *Astronomers: The Biographical Dictionary of Scientists.* New York: Peter Bedrick Books, 1984.

Abell, George, David Morrison, and Sidney C. Wolff. *Realm of the Universe.* Philadelphia: Saunders College Publishing, 1988.

Asimov, Isaac. *Asimov's Biographical Encyclopedia of Science and Technology.* Garden City, N.Y.: Doubleday, 1982.

Bailey, M. E., and D. A. Williams. *Dust in the Universe.* Cambridge, England: Cambridge University Press, 1988.

Barnard, Edward Emerson. *A Photographic Atlas of Selected Regions of the Milky Way.* Washington, D.C.: Carnegie Institution, 1927.

Bartusiak, Marcia F. *Thursday's Universe.* New York: Random House, 1986.

Berry, Arthur. *A Short History of Astronomy.* New York: Dover Publications, 1961.

Bok, Bart J., and Priscilla F. Bok. *The Milky Way.* Cambridge, Mass.: Harvard University Press, 1981.

Burbidge, Geoffrey, David Layzer, and John G. Phillips (eds.):
Annual Review of Astronomy and Astrophysics (Vol. 21). Palo Alto, Calif.: Annual Reviews, 1983.
Annual Review of Astronomy and Astrophysics (Vol. 27). Palo Alto, Calif.: Annual Reviews, 1989.

Chaisson, Eric. *Universe: An Evolutionary Approach to Astronomy.* Englewood Cliffs, N.J.: Prentice-Hall, 1988.

Cornell, James, and John Carr. *Infinite Vistas: New Tools for Astronomy.* New York: Charles Scribner's Sons, 1985.

Daintith, John, Sarah Mitchell, and Elizabeth Tootill. *A Biographical Encyclopedia of Scientists.* New York: Facts on File, 1981.

Ferris, Timothy. *Galaxies.* New York: Stewart, Tabori & Chang, 1982.

Friedlander, Michael W. *Astronomy: From Stonehenge to Quasars.* Englewood Cliffs, N.J.: Prentice-Hall, 1985.

Harwit, Martin. *Cosmic Discovery.* New York: Basic Books, 1981.

Henbest, Nigel, and Michael Martin. *The New Astronomy.* Cambridge, England: Cambridge University Press, 1983.

Hey, J. S. *The Evolution of Radio Astronomy.* London: Eleck Science, 1973.

Hoskin, Michael A. *William Herschel and the Construction of the Heavens.* New York: W. W. Norton, 1963.

Kargon, Robert H. *The Rise of Robert Millikan: Portrait of a Life in American Science.* Ithaca, N.Y.: Cornell University Press, 1982.

Kaufmann, William J., III. *Universe* (2d ed.). New York: W. H. Freeman, 1987.

Lang, Kenneth R., and Owen Gingerich (eds.). *A Source Book in Astronomy and Astrophysics, 1900-1975.* Cambridge, Mass.: Harvard University Press, 1979.

Lynds, Beverly T. (ed.). *Dark Nebulae, Globules, and Protostars.* Tucson: University of Arizona Press, 1971.

Malin, David, and Paul Murdin. *Colours of the Stars.* Cambridge, England: Cambridge University Press, 1984.

Mihalas, Dimitri, and James Binney. *Galactic Astronomy: Structure and Kinematics* (2d ed.). San Francisco: W. H. Freeman, 1981.

Moore, Patrick. *The International Encyclopedia of Astronomy.* New York: Orion Books, 1987.

Morris, Richard. *The Fate of the Universe.* New York: Playboy Press, 1982.

Motz, Lloyd, and Carol Nathanson. *The Constellations.* New York: Doubleday, 1988.

Parker, Barry. *Invisible Matter and the Fate of the Universe.* New York: Plenum Press, 1989.

Pasachoff, Jay M. *Astronomy: From the Earth to the Universe.* Philadelphia: Saunders College Publishing, 1987.

Preiss, Byron, and Andrew Fraknoi (eds.). *The Universe.* New York: Bantam Books, 1987.

Press, Jaques Cattell (ed.). *American Men and Women of Science* (13th ed.). New York: R. R. Bowker, 1976.

Sarazin, Craig L. *X-Ray Emission from Clusters of Galaxies.* Cambridge, England: Cambridge University Press, 1988.

Shu, Frank H. *The Physical Universe: An Introduction to*

Astronomy. Mill Valley, Calif.: University Science Books, 1982.

Silk, Joseph. *The Big Bang.* New York: W. H. Freeman, 1989.

Snow, Theodore P. *Essentials of the Dynamic Universe: An Introduction to Astronomy* (2d ed.). St. Paul, Minn.: West, 1987.

Solomon, P. M., and M. G. Edmunds (eds.). *Giant Molecular Clouds in the Galaxy: Third Gregynog Astrophysics Workshop.* Oxford, England: Pergamon Press, 1980.

Spitzer, Lyman, Jr. *Searching between the Stars.* New Haven, Conn.: Yale University, 1982.

Stars (Voyage Through the Universe series). Alexandria, Va.: Time-Life Books, 1988.

Struve, Otto, and Velta Zebergs. *Astronomy of the 20th Century.* New York: Macmillan, 1962.

Tucker, Wallace H., and Karen Tucker. *The Dark Matter: Contemporary Science's Quest for the Mass Hidden in Our Universe.* New York: William Morrow, 1988.

Verschuur, Gerrit L. *Interstellar Matters: Essays on Curiosity and Astronomical Discovery.* New York: Springer-Verlag, 1989.

Verschuur, Gerrit L., and K. I. Kellermann (eds.). *Galactic and Extragalactic Radio Astronomy* (2d ed.). New York: Springer-Verlag, 1988.

Woerden, Hugo van, III. *Oort and the Universe.* Dordrecht, Netherlands: D. Reidel, 1980.

Wolstencroft, R. D., and W. B. Burton (eds.). *Millimetre and Submillimetre Astronomy.* Dordrecht, Netherlands: Kluwer Academic, 1988.

Zeilik, Michael, and John Gaustad. *Astronomy: The Cosmic Perspective.* New York: Harper & Row, 1983.

Zeilik, Michael, and Elske v. P. Smith. *Introductory Astronomy and Astrophysics* (2d ed.). Philadelphia: Saunders College Publishing, 1987.

Periodicals

Adams, Walter S. "Some Results with the Coudé Spectrograph of the Mount Wilson Observatory." *Publications of the Astronomical Society of the Pacific,* December 1941.

Allamandola, L. J., S. A. Sandford, and B. Wopenka. "Interstellar Polycyclic Aromatic Hydrocarbons and Carbon in Interplanetary Dust Particles and Meteorites." *Science,* July 3, 1987.

Asimov, Isaac. "Science: Out of the Everywhere." *Magazine of Fantasy & Science Fiction,* November 1988.

Balick, Bruce. "The Shaping of Planetary Nebulae." *Sky & Telescope,* February 1987.

Bally, John. "Interstellar Molecular Clouds." *Science,* April 11, 1986.

Barnard, E. E. "On the Dark Markings of the Sky." *Astrophysical Journal,* January 1919.

Barrett, Alan H. "Radio Signals from Hydroxyl Radicals." *Scientific American,* December 1968.

Bartusiak, Marcia. "Signposts in the Sky." *Discover,* July 1981.

Blitz, Leo. "Giant Molecular-Cloud Complexes in the Galaxy." *Scientific American,* April 1982.

"Carnegie Institution's Mt. Wilson Observatory Makes Astronomical History." *Life,* November 8, 1937.

Carruthers, George R. "Rocket Observation of Interstellar Molecular Hydrogen." *Astrophysical Journal,* August 1970.

Chandrasekhar, S., and E. Fermi. "Magnetic Fields in Spiral Arms." *Astrophysical Journal,* July 1953.

Dame, Thomas M. "The Molecular Milky Way." *Sky & Telescope,* July 1988.

Darling, David. "Breezes, Bangs, & Blowouts: Stellar Evolution through Mass Loss." *Astronomy,* November 1985.

Davis, Leverett, Jr., and Jesse L. Greenstein. "The Polarization of Starlight by Aligned Dust Grains." *Astrophysical Journal,* September 1951.

De Boer, Klaas S., and Blair D. Savage. "The Coronas of Galaxies." *Scientific American,* August 1982.

Dickinson, Dale F. "Cosmic Masers." *Scientific American,* June 1978.

Eddington, Arthur S. "Bakerian Lecture—Diffuse Matter in Interstellar Space." *Proceedings of the Royal Society: Mathematical and Physical Sciences,* Series A, Vol. 3, no. A759, July 2, 1926.

"520 Light-Years Away, a Star Is Born." *Discover,* November 1986.

"Formaldehyde in Space." *Scientific American,* May 1969.

Frost, Edwin B. "Spectrographic Notes." *Astrophysical Journal,* April 1909.

Graham, David. "The Search for the Invisible Universe." *Technology Review,* October 1988.

Greenberg, J. Mayo. "The Structure and Evolution of Interstellar Grains." *Scientific American,* June 1984.

Gusten, R., et al. "Aperture Synthesis Observations of the Circumnuclear Ring in the Galactic Center." *Astrophysical Journal,* July 1, 1987.

Hall, John S. "V. M. Slipher's Trailblazing Career." *Sky & Telescope,* February 1970.

Heger, Mary Lea. "The Occurrence of Stationary D Lines of Sodium in the Spectroscopic Binaries, β Scorpii and δ Orionis." *Lick Observatory Bulletin,* 1919, Vol. 10, no. 326.

Helfand, David J. "Fleet Messengers from the Cosmos." *Sky & Telescope,* March 1988.

Hills, Richard E., et al. "The Hat Creek Millimeter-Wave Interferometer." *Proceedings of the IEEE,* September 1973.

Kaler, James B. "Origins of the Spectral Sequence." *Sky & Telescope,* February 1986.

Killian, Anita. "Galactic Center Update." *Sky & Telescope,* March 1986.

Kunzig, Robert. "Stardust Memories: Kiss of Life." *Discover,* March 1988.

Lada, Charles J., et al. "Molecular Clouds in the Vicinity of W3, W4, and W5." *Astrophysical Journal,* November 15, 1978.

McCray, Richard, and Minas Kafatos. "Supershells and Propagating Star Formation." *Astrophysical Journal,* June 1, 1987.

McKellar, Andrew. "Evidence for the Molecular Origin of Some Hitherto Unidentified Interstellar Lines." *Publications of the Astronomical Society of the Pacific,* June 1940.

Maddalena, Ronald J., et al. "The Large System of Molecular Clouds in Orion and Monoceros." *Astrophysical Journal,* April 1, 1986.

Malin, David. "In the Shadow of the Horsehead." *Sky & Telescope,* September 1987.

Melott, Adrian. "The Invisible Universe." *Astronomy,*

May 1981.

Mitchell, S. A. "Edward Emerson Barnard, 1857-1928." *Observatory*, May 1928.

Mumford, George S. "The Legacy of E. E. Barnard." *Sky & Telescope*, July 1987.

Norman, Colin A., and Satoru Ikeuchi. "The Disk-Halo Interaction: Superbubbles and the Structure of the Interstellar Medium." *Astrophysical Journal*, October 1, 1989.

Norris, Ray. "Cosmic Masers." *Sky & Telescope*, March 1986.

"PAH's: A New Breed of Interstellar Matter." *Sky & Telescope*, March 1990.

Parker, Barry. "The Mysterious Dark Matter." *Stardate*, December 1989.

Payne-Gaposchkin, Cecilia. "Otto Struve as an Astrophysicist." *Sky & Telescope*, June 1963.

Plambeck, R. L., et al. "Kinematics of Orion-KL: Aperture Synthesis Maps of 86 GHzSO Emission." *Astrophysical Journal*, August 15, 1982.

Plaskett, J. S. "The H and K Lines of Calcium in O-Type Stars." *Monthly Notices of the Royal Astronomical Society*, December 14, 1923.

"The Rain in Space." *Scientific American*, April 1969.

Rank, D. M., C. H. Townes, and W. J. Welch. "Interstellar Molecules and Dense Clouds." *Science*, December 10, 1971.

"Research on Maser-Laser Principle Wins Nobel Prize in Physics." *Science*, November 13, 1964.

Schneider, Stephen E., and Yervant Terzian. "Between the Galaxies." *American Scientist*, November-December 1984.

Scoville, Nick, and Judith S. Young. "Molecular Clouds, Star Formation and Galactic Structure." *Scientific American*, April 1984.

Seeley, D., and R. Berendzen. "The Development of Research in Interstellar Absorption, c.1900-1930." *Journal of the History of Astronomy*, 1972, pp. 52-64.

Shore, Lys Ann, and Steven N. Shore. "The Chaotic Material between the Stars." *Astronomy*, June 1988.

"Sites of Star Formation." *Science*, May 4, 1990.

Solomon, Philip M. "Interstellar Molecules." *Physics Today*, March 1973.

Spitzer, Lyman, Jr. "Dreams, Stars, and Electrons." *Annual Review of Astronomy and Astrophysics*, 1989.

"Split Award." *Time*, November 6, 1964.

Townes, Charles H., and Reinhard Genzel. "What Is Happening at the Center of Our Galaxy?" *Scientific American*, April 1990.

Trimble, Virginia. "Existence and Nature of Dark Matter in the Universe." *Annual Review of Astronomy and Astrophysics*, 1987.

Turner, Barry E.:
"Interstellar Molecules." *Scientific American*, March 1973.
"Recent Progress in Astrochemistry." *Space Science Reviews*, 1989, Vol. 51, pp. 235-337.

Vaughan, Christopher. "Tracking an Elusive Carbon." *Science News*, January 28, 1989.

Verschuur, Gerrit L.:
"Molecules between the Stars." *Mercury*, May-June 1987.
"Recent Measurements of the Zeeman Effect at 21-Centimeter Wavelength." *Astrophysical Journal*, May 1, 1971.

Waldrop, M. Mitchell. "Stellar Nurseries." *Science '83*, May 1983.

"Water Molecules in Space." *New York Times*, November 17, 1963.

Weaver, Harold, et al. "Observations of a Strong Unidentified Microwave Line and of Emission from the OH Molecule." *Nature*, October 2, 1965.

Weinreb, S., et al. "Radio Observations of OH in the Interstellar Medium." *Nature*, November 30, 1963.

Wilson, R. W., K. B. Jefferts, and A. A. Penzias. "Carbon Monoxide in the Orion Nebula." *Astrophysical Journal*, July 1970.

Other Sources

Allamandola, Louis. "Summary of: Interstellar Polycyclic Aromatic Hydrocarbons." News release on paper no. 516. Moffett Field, Calif.: NASA Ames Research Center.

Carlstrom, John E. "Aperture Synthesis Maps of HCN, HCO, CO, and 3 mm Continuum toward M82." RAL Preprint no. 123. Berkeley: Radio Astronomy Laboratory, University of California, October 1987.

Frost, Edwin B. "Biographical Memoir: Edward Emerson Barnard: 1857-1923." Memoir no. 14. Washington, D.C.: National Academy of Sciences, 1924.

Gammon, Richard H. "Chemistry between the Stars." Curriculum project. Washington, D.C.: NASA, September 1976.

INDEX

ACKNOWLEDGMENTS

The editors wish to thank Louis Allamandola, NASA Ames Research Center, Moffett Field, Calif.; Thomas M. Bania, Boston University; John Carlstrom, California Institute of Technology; S. Chandrasekhar, Laboratory for Astrophysics and Space Research, Chicago; Donna Coletti, Center for Astrophysics, Cambridge, Mass.; David De Young, Kitt Peak National Observatory, Tucson; Richard Dreiser, University of Chicago, Williams Bay, Wis.; Esther Ferington, Time-Life Books, Alexandria, Va.; Herbert Gursky, Naval Research Laboratory, Washington, D.C.; Eric Herbst, Duke University; Charles J. Lada, University of Arizona, Tucson; Ronald J. Maddalena, National Radio Astronomy Observatory, Green Bank, W.Va.; Donald Osterbrock, Princeton University; Stephen Schneider, University of Massachusetts, Amherst; Richard Smalley, Rice University; Lyman Spitzer, Princeton University; Michael S. Turner, Fermi National Accelerator Laboratory, Batavia, Ill.; Marie-Josée Vin, Observatoire de Haute-Provence, France; Sander Weinreb, Martin Marietta Laboratories, Baltimore; Arthur Wolfe, University of California, San Francisco; Alwyn Wootten, National Radio Astronomy Observatory, Charlottesville, Va.; Melvyn Wright, University of California, Berkeley.

PICTURE CREDITS

Time-Life Books
is a wholly owned subsidiary of
THE TIME INC. BOOK COMPANY

President and Chief Executive Officer:
Kelso F. Sutton
President, Time Inc. Books Direct:
Christopher T. Linen

TIME-LIFE BOOKS INC.
EDITOR: George Constable
Director of Design: Louis Klein
Director of Editorial Resources: Phyllis K. Wise
Director of Photography and Research:
John Conrad Weiser

PRESIDENT: John M. Fahey, Jr.
Senior Vice Presidents: Robert M. DeSena, Paul R.
Stewart, Curtis G. Viebranz, Joseph J. Ward
Vice Presidents: Stephen L. Bair, Bonita L.
Boezeman, Mary P. Donohoe, Stephen L.
Goldstein, Andrew P. Kaplan, Trevor Lunn,
Susan J. Maruyama, Robert H. Smith
New Product Development: Trevor Lunn,
Donia Ann Steele
Supervisor of Quality Control: James King

PUBLISHER: Joseph J. Ward

Editorial Operations
Production: Celia Beattie
Library: Louise D. Forstall

Computer Composition: Deborah G. Tait
(Manager), Monika D. Thayer, Janet Barnes
Syring, Lillian Daniels

Correspondents: Elisabeth Kraemer-Singh (Bonn),
Christine Hinze (London), Christina Lieberman
(New York), Maria Vincenza Aloisi (Paris), Ann
Natanson (Rome). Valuable assistance was also
provided by Elizabeth Brown (New York), John
Dunn (Melbourne), Mary Johnson (Stockholm),
Wibo Van de Linde (Amsterdam).

VOYAGE THROUGH THE UNIVERSE

SERIES DIRECTOR: Roberta Conlan
Series Administrator: Susan Stuck

Editorial Staff for *Between the Stars*
Designers: Robert K. Herndon (principal),
Dale Pollekoff
Associate Editor: Kristin Baker Hanneman
(pictures)
Text Editors: Allan Fallow, Stephen Hyslop,
Robert M. S. Somerville
Researchers: Mark Galan, Karin Kinney,
Elizabeth Thompson
Assistant Designer: Barbara M. Sheppard
Copy Coordinators: Darcie Conner Johnston,
Juli Duncan
Picture Coordinators: Barry Anthony,
Jennifer Iker
Editorial Assistant: Katie Mahaffey

Special Contributors: Deborah Byrd, George
Daniels, James Dawson, Jane Ferrell, Stephen
Hart, Jeff Kanipe, Frank Kendig, John Langone,
Gina Maranto, Eliot Marshall, Barry Parker,
Chuck Smith, Larry Thompson, Elizabeth Ward,
Mark Washburn (text); Vilasini Balakrishnan,
Craig Chapin, Nancy L. Connors, Edward Dixon,
Jocelyn G. Lindsay, Jacqueline Shaffer, Roberta
Yared (research); Barbara L. Klein (index).

CONSULTANTS

BRUCE BALICK, a professor of astronomy at the
University of Washington, examines the processes
by which dying stars eject material that seeds new
generations of stars.

CHRIS BLADES is an astrophysicist at the Space
Telescope Science Institute in Baltimore, Maryland,
specializing in the study of interstellar and quasar
absorption lines.

LEO BLITZ, a radio astronomer and specialist in
interstellar matter, heads the Laboratory for
Millimeter-Wave Astronomy at the University of
Maryland at College Park.

JOEL N. BREGMAN is an astronomy professor at
the University of Michigan, where he investigates
the interstellar medium in galaxies. His other interests include active galactic nuclei and theoretical astrophysics.

MARTIN COHEN is affiliated with the Radio Astronomy Laboratory at the University of California,
Berkeley. For the past twenty years, he has studied
star formation and the nature of interstellar and
circumstellar dust.

EUGENE J. DE GEUS is a research associate in the
Astronomy Program at the University of Maryland,
where he explores the interaction of massive stars
with the interstellar medium.

JAMES B. KALER, an expert on spectroscopy, teaches stellar astronomy at the University of Illinois.

JOHN MATHIS is a professor of astronomy at the
University of Wisconsin whose recent research
probes the nature and size distribution of interstellar dust grains.

RICHARD MUSHOTZKY, an x-ray astronomer at
NASA Goddard Space Flight Center in Greenbelt,
Maryland, specializes in the study of active galaxies
and clusters of galaxies.

MARK J. REID is a radio astronomer at the Harvard-
Smithsonian Center for Astrophysics. His areas of
research include the study of astronomical sources
of maser emission and a technique called Very Long
Baseline Interferometry (VLBI).

STEVEN SHORE, a theoretical astrophysicist, is a
member of the High Resolution Spectrograph Science team at NASA Goddard Space Flight Center.

PATRICK THADDEUS, a physicist who teaches at
Harvard University, has worked in radio astronomy
for many years. He developed the minitelescope
used to map molecular clouds.

SCOTT TREMAINE of the Canadian Institute for
Theoretical Astrophysics at the University of To-
ronto specializes in galaxy and Solar System dynamics.

BARRY E. TURNER, a senior staff scientist at the
National Radio Astronomy Observatory in Charlottesville, Virginia, conducts research in the chemistry and physics of the interstellar medium.

GERRIT L. VERSCHUUR, an astronomer who specializes in the study of interstellar hydrogen, teaches astronomy at the University of Maryland and
publishes widely on the subject of astronomy.

**Library of Congress Cataloging in
Publication Data**
Between the stars / by the editors of Time-Life
Books.
p. cm. (Voyage through the universe).
Bibliography: p.
Includes index.
ISBN 0-8094-6895-6.
ISBN 0-8094-6896-4 (lib. bdg.).
1. Astronomy—History.
I. Time-Life Books. II. Series.
QB28.B47 1990
520'.9—dc20   90-11028 CIP

For information on and a full description of
any of the Time-Life Books series, please call
1-800-621-7026 or write:
Reader Information
Time-Life Customer Service
P.O. Box C-32068
Richmond, Virginia 23261-2068

Time-Life Books Inc. offers a wide range of fine
recordings, including a *Rock 'n' Roll Era* series.
For subscription information, call 1-800-621-7026
or write Time-Life Music, P.O. Box C-32068, Richmond, Virginia 23261-2068.

Earth: diameter 7,926 miles

Neptune: diameter 30,700 miles

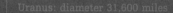

Uranus: diameter 31,600 miles

Red supergiant: diameter 400 million miles

Solar System: diameter 7.5 billion miles

Globular cluster: diameter 2×10^{14} miles

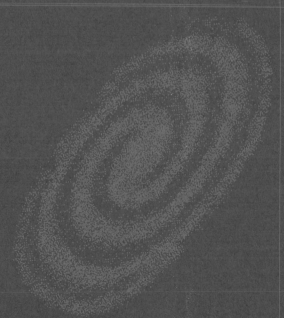

Milky Way: diameter 100,000 light-years

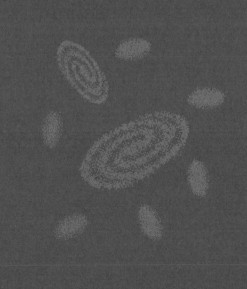

Local Group of galaxies:
6 million light-years across

Largest double radio source:
length 17 million light-years